Medical Microbiology
and Infection
at a Glance

Medical Microbiology and Infection at a Glance

STEPHEN H. GILLESPIE
MD, FRCP (Edin), FRCPath
Professor of Medical Microbiology
Royal Free and University College Medical School
(Royal Free Campus)
University College London
London

KATHLEEN B. BAMFORD
MD, FRCPath
Senior Lecturer in Medical Microbiology
Imperial College School of Medicine
Hammersmith Hospital
London

Edited by
JANET P. GILLESPIE MB, MRCGP
General Practitioner

**Blackwell
Science**

© 2000 by
Blackwell Science Ltd
Editorial Offices:
Osney Mead, Oxford OX2 0EL
25 John Street, London WC1N 2BL
23 Ainslie Place, Edinburgh EH3 6AJ
350 Main Street, Malden
  MA 02148 5018, USA
54 University Street, Carlton
  Victoria 3053, Australia
10, rue Casimiar Delavigne
  75006 Paris, France

Other Editorial Offices:
Blackwell Wissenschafts-Verlag GmbH
Kurfürstendamm 57
10707 Berlin, Germany

Blackwell Science KK
MG Kodenmacho Building
7–10 Kodenmacho Nihombashi
Chuo-ku, Tokyo 104, Japan

First published 2000
Reprinted 2000

Set by Excel Typesetters Co., Hong Kong
Printed and bound in Great Britain at
the Alden Press, Oxford and Northampton

The Blackwell Science logo is a
trade mark of Blackwell Science Ltd,
registered at the United Kingdom
Trade Marks Registry

A catalogue record for this title
is available from the British Library

ISBN 0-632-05026-8

Library of Congress
Cataloging-in-publication Data

Gillespie, S. H.
      Medical microbiology and infection
at a glance /
   Stephen H. Gillespie,
   Kathleen B. Bamford;
   Edited by
   Janet P. Gillespie.
      p.   cm.
   Includes index.
   ISBN 0-632-05026-8
   1. Medical microbiology.
   I. Bamford, Kathleen.
   II. Title.
   [DNLM: 1. Microbiology.
   2. Communicable Diseases.
QW 4 G478m 2000]
QR46.G47 2000
616′.01 — dc21
DNLM/DLC
for Library of Congress   99-26884
                                       CIP

DISTRIBUTORS

  Marston Book Services Ltd
  PO Box 269
  Abingdon, Oxon OX14 4YN
  (Orders: Tel: 01235 465500
              Fax: 01235 465555)

USA
  Blackwell Science, Inc.
  Commerce Place
  350 Main Street
  Malden, MA 02148 5018
  (Orders: Tel: 800 759 6102
                 781 388 8250
           Fax: 781 388 8255)

Canada
  Login Brothers Book Company
  324 Saulteaux Crescent
  Winnipeg, Manitoba R3J 3T2
  (Orders: Tel: 204 837-2987)

Australia
  Blackwell Science Pty Ltd
  54 University Street
  Carlton, Victoria 3053
  (Orders: Tel: 3 9347 0300
              Fax: 3 9347 5001)

For further information on
Blackwell Science, visit our website:
www.blackwell-science.com

Contents

Preface

This book is written for medical students and doctors who are seeking a brief summary of microbiology and infectious diseases. It should prove useful to those embarking on a course of study and assist those preparing for professional examinations.

Chapters are divided into concepts, the main human pathogens and the infectious syndromes. This broadly reflects the pattern of teaching in many medical schools.

Microbiology is a rapidly growing and changing subject: new organisms are constantly being identified and our understanding of the pathogenic potential of recognized pathogens is being expanded. In addition the taxonomists keep changing the names of familiar friends to add to the confusion. Despite this, there are clear fundamental facts and principles that form a firm foundation of knowledge on which to build throughout a professional career. It is these that this book strives to encapsulate.

Each chapter contains a diagram which illustrates core knowledge. The associated text offers further insights and details where necessary.

Irrespective of a doctor's specialty, diligent study of microbiology provides the basis for sound professional judgement, giving the clinician confidence and benefiting patients for years to come.

The authors gratefully acknowledge the editorial work of Dr Janet Gillespie who has reminded the authors of practice in a community setting. They are also grateful to Dr Deenan Pillay for his critical reading of the virology sections.

Stephen Gillespie & Kathleen Bamford
London, 2000

1 Structure and classification of bacteria

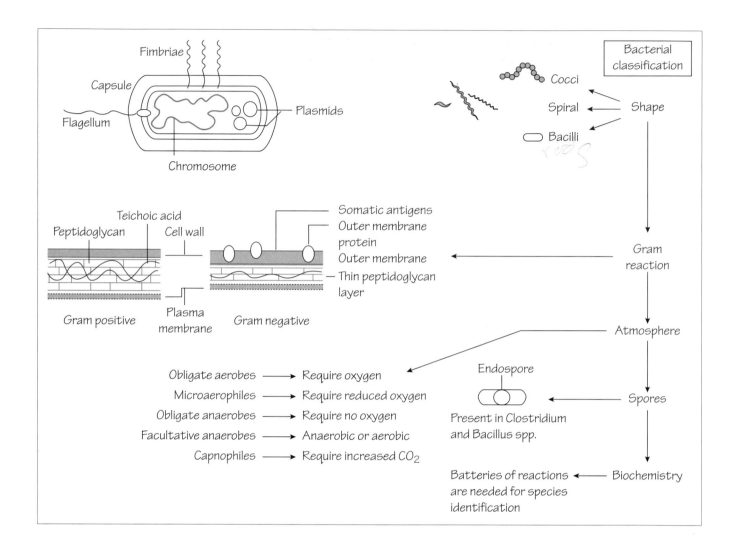

Bacterial organelles

Bacteria possess a rigid cell wall that is responsible for maintaining their shape and protecting the cell from differences in osmotic tension between the inside and outside environment. Gram-positive walls have a thick peptidoglycan layer and a cell membrane, whereas Gram-negative cells have three layers with an inner and outer membrane and a thinner peptidoglycan layer. Mycobacteria cell wall has a high proportion of lipid, including immunoreactive antigens. Bacterial shape depends on cell wall structure: cocci are spherical, bacilli are long and thin, with coccobacilli in between, and spiral shapes of different wavelength. Shape is used in bacterial classification. Outside the cell wall the bacterium may express important antigens and structures.

- *Capsule*: a loose polysaccharide structure protecting it from phagocytosis and desiccation.

- *Lipopolysaccharide*: found in Gram-negative bacteria protecting against complement-mediated lysis and a potent stimulator of cytokine release.
- *Fimbriae or pili*: specialized thin projections that aid adhesion to host cells and colonization. Uropathogenic *Escherichia coli* have specialized fimbriae (P fimbriae) that bind to mannose receptors on ureteric epithelial cells. Fimbrial antigens are often immunogenic but vary between strains so that repeated infections may occur, e.g. *Neisseria gonorrhoeae*.
- *Flagella*: bacterial organs of locomotion enabling organisms to find sources of nutrition and penetrate host mucus. Flagella can be single or multiple, at one end of the cell (polar) or at many points (peritrichous). In some species, e.g. *Treponema*, the flagella are firmly fixed within the bacterial cell wall.
- *Slime*: polysaccharide material secreted by some bacteria

growing in biofilms and protects the organism against immune attack and eradication by antibiotics.

- *Spores*: metabolically inert form triggered by adverse environmental conditions; adapted for long-term survival, it allows regrowth when conditions allow.

Bacteria are prokaryotes, i.e. they have a single chromosome and lack a nucleus. To pack the chromosome inside the cell the DNA is coiled and supercoiled; a process mediated by the DNA gyrase enzyme system. Bacterial ribosomes differ from eukaryotic, e.g. size difference makes this a target for antibacterial therapy. Because bacteria lack membrane-bound organelles metabolic processes must take place in the cytoplasm.

Bacteria also contain accessory DNA in the form of plasmids. The role of plasmids in antimicrobial resistance is discussed in more detail in Chapter 6.

Classification of bacteria

The purpose of classification of microorganisms is to define the pathogenic potential: for example, a *Staphylococcus aureus* isolated from blood is more likely to be acting as a pathogen than *Staphylococcus epidermidis* from the same site. Some bacteria have the capacity to spread widely in the community and cause serious disease, e.g. *Corynebacterium diphtheriae* and *Vibrio cholerae*. Bacteria are identified, or speciated, by using a series of physical characteristics. Some of these are listed below.

- Gram reaction: Gram-positive and -negative bacteria respond to different antibiotics. Other bacteria, e.g. *Mycobacteria*, may need special staining techniques.
- Cell shape (cocci, bacilli or spirals).
- Presence and shape of an endospore and its position in the bacterial cell (terminal, subterminal or central).
- Atmospheric preference: aerobic organisms require oxygen; anaerobic require an atmosphere with very little or no oxygen. Organisms that grow in either atmosphere are known as facultative anaerobes. Microaerophiles prefer a reduced oxygen tension; capnophiles prefer increased carbon dioxide.
- Requirement for special media or intracellular growth.

More detailed biochemical, antigenic and molecular tests are performed to identify the species of organisms (see Chapter 4).

Medically important groups of bacteria

Gram-positive cocci

Gram-positive cocci are divided into two main groups: the staphylococci (catalase-positive), e.g. the major pathogen *Staphylococcus aureus*, and the streptococci (catalase-negative). The latter include *Streptococcus pyogenes*, an agent of sore throat and rheumatic fever, and *Streptococcus agalactiae*, a cause of neonatal meningitis and pneumonia (see Chapter 11).

Gram-negative cocci

Gram-negative cocci include the pathogenic *Neisseria meningitidis*, an important cause of meningitis and septicaemia, and *N. gonorrhoeae*, an agent of sexually transmitted urethritis (gonorrhoea).

Cocco-bacilli

These Gram-negative organisms include the respiratory pathogens *Haemophilus* and *Bordetella*, and zoonotic agents, such as *Brucella* and *Pasteurella* (see Chapter 16).

Gram-positive bacilli

Gram-positive bacilli are divided into sporing and non-sporing; the sporing are subdivided between those that are aerobic (*Bacillus*: see Chapter 12) and those that are anaerobic (*Clostridium*: see Chapter 14). Pathogens include *Bacillus anthracis* and those of gas gangrene, tetanus, pseudomembranous colitis and botulism. Non-sporing pathogens include *Listeria* and corynebacteria (see Chapter 12).

Gram-negative bacilli

Facultative Gram-negative bacilli, including the family Enterobacteriaceae, form part of the normal flora of humans and animals and can be found in the environment. They include many pathogenic genera: *Salmonella, Shigella, Escherichia, Proteus* and *Yersinia* (see Chapter 18). *Pseudomonas* is an aerobic environmental saprophyte naturally resistant to antibiotics that has come to be an important pathogen in the hospital environment (see Chapter 20). *Legionella* is another environmental species that lives in water but causes human infection if conditions allow (see Chapter 20).

Spiral bacteria

Gastrointestinal pathogens include the small spiral *Helicobacter* that colonizes the stomach, leading to gastric and duodenal ulcer and gastric cancer, and *Campylobacter* spp. that cause acute diarrhoea. The *Borrelia* give rise to relapsing fever (*B. duttoni* and *B. recurrentis*) and to a chronic disease of the skin joints and central nervous system, Lyme disease (*B. burgdorferi*). The *Leptospira* are zoonotic agents causing an acute meningitis syndrome that may be accompanied by renal failure and hepatitis. The *Treponema* include the causative agent of syphilis (*T. pallidum*).

Rickettsia, Chlamydia and *Mycoplasma*

Of these, only *Mycoplasma* can be isolated on artificial media: the others require isolation in cell culture, or diagnosis by molecular or serological techniques.

2 Innate immunity and normal flora

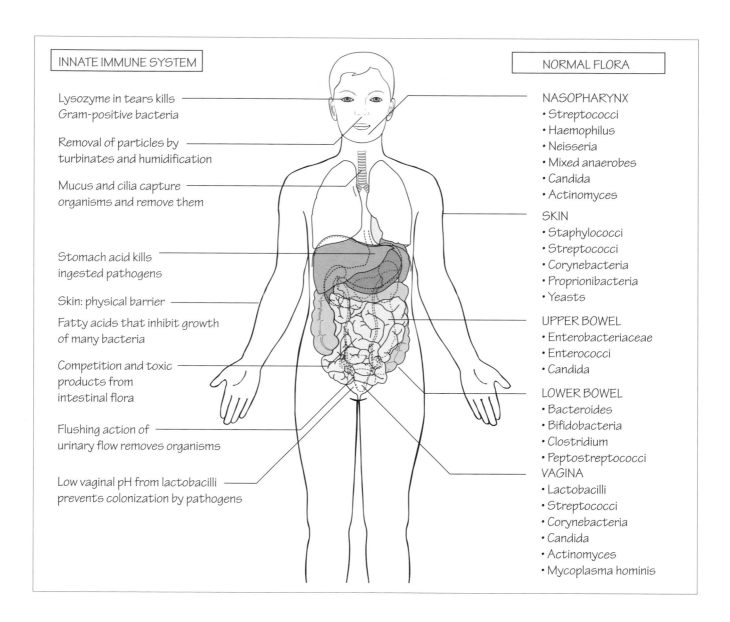

INNATE IMMUNE SYSTEM

Lysozyme in tears kills
Gram-positive bacteria

Removal of particles by
turbinates and humidification

Mucus and cilia capture
organisms and remove them

Stomach acid kills
ingested pathogens

Skin: physical barrier

Fatty acids that inhibit growth
of many bacteria

Competition and toxic
products from
intestinal flora

Flushing action of
urinary flow removes organisms

Low vaginal pH from lactobacilli
prevents colonization by pathogens

NORMAL FLORA

NASOPHARYNX
• Streptococci
• Haemophilus
• Neisseria
• Mixed anaerobes
• Candida
• Actinomyces

SKIN
• Staphylococci
• Streptococci
• Corynebacteria
• Proprionibacteria
• Yeasts

UPPER BOWEL
• Enterobacteriaceae
• Enterococci
• Candida

LOWER BOWEL
• Bacteroides
• Bifidobacteria
• Clostridium
• Peptostreptococci

VAGINA
• Lactobacilli
• Streptococci
• Corynebacteria
• Candida
• Actinomyces
• Mycoplasma hominis

Normal flora

The body contains a huge number of resident microorganisms: in fact prokaryotic (bacterial) cells outnumber human cells. The normal flora protects the body by competing with pathogens for colonization sites, and producing antibiotic substances (bacteriocins) that suppress competing organisms. Anaerobes produce toxic metabolic products and free fatty acids that inhibit other organisms. In the female genital tract, *Lactobacilli* produce lactic acid that lowers the pH preventing colonization by pathogens.

Antibiotics suppress normal flora allowing colonization and infection by naturally resistant organisms, e.g. *Candida albicans*. The infective dose of *Salmonella typhi* is lowered by concomitant antibiotic use. Antibiotics may upset the balance between organisms of the normal flora allowing one to predominate, e.g. *Clostridium difficile* infection.

Skin

The skin provides a physical barrier to invasion, secreting sebum and fatty acids that inhibit bacterial growth. Many organisms have evolved mechanisms to penetrate the skin, whether via the bite of a vector, e.g. dengue fever following the bite of *Aedes aegypti*, or invasion through intact skin, e.g. *Leptospira* and *Treponema*. Some organisms colonize mucosal surfaces and use this route to gain access to the body.

Loss of skin integrity by intravenous cannulation, or medical or non-medical injection, can transmit blood-borne viruses such as hepatitis B or HIV. Diseases of the skin, e.g. eczema or burns, permit colonization and invasion by pathogens, e.g. *Streptococcus pyogenes*.

Mucociliary clearance mechanism

In the respiratory tract air is humidified and warmed by passage over the turbinate bones and through the nasal sinuses. Particles settle on the sticky mucus of the respiratory epithelium: the debris is transported by the cilial 'conveyor belt' to the oropharynx where it is swallowed. This efficient system only allows particles less than 5 μm diameter to reach the alveoli: the respiratory tract is effectively sterile below the carina.

Secreted antibacterial compounds

Mucus contains polysaccharides of similar antigenic structure to the underlying mucosal surface: organisms bind to the mucus and are removed. The body secretes antibacterial compounds, e.g. lysozyme in tears degrades Gram-positive bacterial peptidoglycan; lactoferrin in breast milk binds iron inhibiting bacterial growth. Lactoperoxidase, a leucocyte enzyme, produces superoxide radicals that are toxic to microorganisms.

Urinary flushing

The urinary tract is protected by the flushing action of urinary flow and, except near the urethral meatus, is sterile. Obstruction by stones, tumours, benign prostatic hypertrophy, or by scarring of the urethra or bladder, may cause reduction of the urinary flow and stasis with subsequent bacterial urinary infection.

Phagocytes

Neutrophils and macrophages ingest particles including bacteria, viruses and fungi. Opsonins, e.g. complement and antibody, may enhance phagocytic ability, e.g. *Streptococcus pneumoniae* are not phagocytosed unless their capsule is coated with an anticapsular antibody. The action of macrophages in the reticuloendothelial system is essential for resistance to many bacterial and protozoan pathogens, e.g. *S. pneumoniae* and malaria. Congenital deficiency in neutrophil function leads to chronic pyogenic infections, recurrent chest infections and bronchiectasis. Following splenectomy, patients have defective macrophage function and diminished ability to remove organisms from the blood.

Complement and other plasma proteins

Complement is a system of plasma proteins that collaborate to resist bacterial infection. The complement cascade is activated by antigen–antibody binding (the classical pathway) or by direct interaction with bacterial cell wall components (the alternative pathway). The products of both processes attract phagocytes to the site of infection (chemotaxis), activate phagocytes, cause vasodilatation and stimulate phagocytosis of bacteria (opsonization). The final three components of the cascade form a 'membrane attack complex' that can lyse Gram-negative bacteria. Complement deficiencies render patients susceptible to acute pyogenic infections especially with *Neisseria meningitidis*, *N. gonorrhoeae* and *Streptococcus pneumoniae*.

Transferrin is a transport vehicle for iron, limiting the amount of iron available to invading microorganisms. Other acute phase proteins are directly antibacterial, e.g. mannose-binding protein or C-reactive protein, which bind to bacteria and activate complement.

Table 2.1 The innate immune system, structure and deficiency

Component	Compromise	Consequence
Normal flora		
Pharynx	Antibiotics	Oral thrush
Intestine	Antibiotics	Pseudomembranous colitis; colonization with antibiotic-resistant organisms
Vagina	Antibiotics	Vaginal thrush
Skin	Burns, vectors	Cutaneous bacterial infection, infection with pathogenic viruses, bacteria, protozoa and metazoa
Turbinates and mucociliary clearance	Kartagener's syndrome, Cystic fibrosis, Bronchiectasis	Chronic bacterial infection
Lysozyme in tears	Sjögren's syndrome	Ocular infection
Urinary flushing	Obstruction	Recurrent urinary infection
Phagocytes, Neutrophils, Macrophages	Congenital, Iatrogenic, Infective	Chronic pyogenic infection, increased susceptibility to bacterial infection
Complement	Congenital deficiency	Increased susceptibility to bacterial infection especially *Neisseria* and *Streptococcus pneumoniae*

3 Pathogenicity and transmission of microorganisms

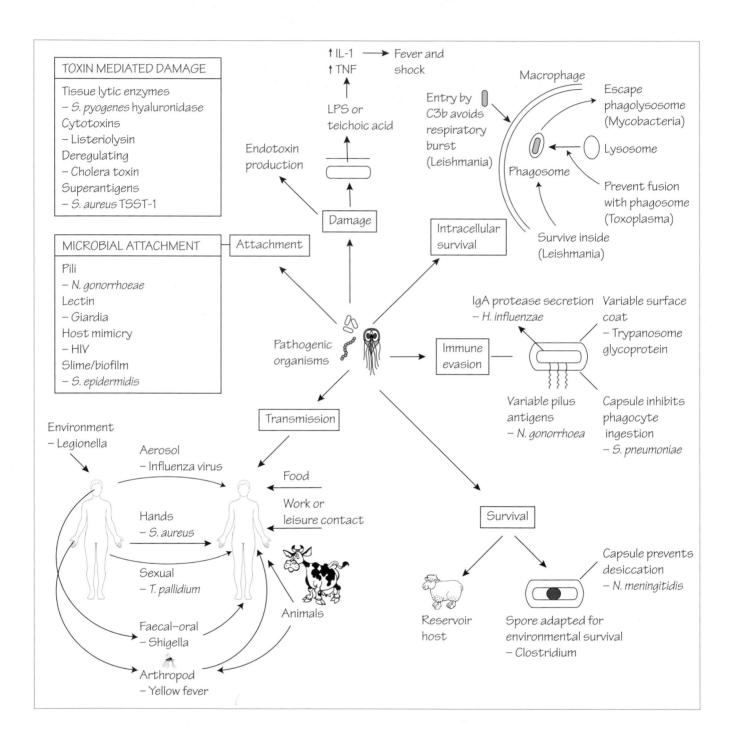

Definitions

The normal host is *colonized* by bacteria and protozoa that do not cause disease. An *infection* occurs when invading microorganisms cause ill-health. An organism capable of causing an infection is a *pathogen*: one which forms part of the normal flora is a *commensal*. *Pathogenicity* is the capa-

city to cause disease, whereas *virulence* is the ability to cause serious disease. The main pathogenicity determinant of *Streptococcus pneumoniae* is the capsule without which it cannot cause disease. Some capsular types cause more serious disease: they alter the virulence. The term *parasite* is often used to describe protozoan and metazoan organisms but this is confusing as these organisms are either pathogens or commensals.

Sources of infection

Obligate pathogens are always associated with disease. Organisms of the normal flora only invade under certain circumstances, causing *endogenous* infection. Animal pathogens may spread to humans by contact or in food: these infections are called *zoonoses*. Humans can become infected from organisms in the inanimate environment, e.g. *Legionella* or *Clostridium tetani*.

Alteration of the environment changes the risk of diseases. Zoonoses are encouraged by intensive farming methods. Feeding ruminant offal to cattle resulted in an epidemic of bovine spongiform encephalopathy. Poorly maintained air-conditioning cooling towers can be a source of *Legionella pneumophila*. Changes in the host alter the risk of disease: surgery and intravenous cannulation favour invasion with organisms of the normal flora and immuno-suppressive therapy makes patients susceptible to opportunists of low virulence.

Microorganisms have developed complex life-cycles to facilitate transmission and survival. Organisms excreted in faeces spread to other hosts by ingestion: the faeco–oral route. Others have a life-cycle stage inside an insect *vector* which transmits the disease by biting. Humans can become infected as an *accidental* host when they substitute for an animal in a life-cycle, e.g. hydatid disease (see Chapter 47).

Survival and transmission

Organisms must survive in the environment. Spores are small structures with a tough coat and a low metabolic rate which enable bacteria to survive for many years. Helminth eggs have a tough shell adapted for survival in the environment. Transmission is favoured when an organism is able to survive in a host which then acts as a *reservoir* of infection.

Microorganisms are propelled out of the nose and mouth in a sneeze and can remain suspended in the air on droplet nuclei (5 μm). Infection may occur when these are inhaled by another person and are carried to the alveoli. Respiratory infections, e.g. influenza, are transmitted this way, as are others which affect other organs, e.g. *Neisseria meningitidis*.

Food and water contain pathogens that may infect the intestinal tract, e.g. *Salmonella*. Toxoplasmosis: cysticercosis, which principally affects other organs, can infect by this route.

Leptospira, *Treponema* and *Schistosoma* have evolved specific mechanisms enabling them to invade intact skin. Injections and blood transfusions bypass the skin allowing the transmission of HIV. Skin organisms, e.g. *Staphylococcus epidermidis*, can invade the body via indwelling venous cannulae. Insects which feed on blood may transmit pathogens: anophelene mosquitoes transmit malaria. Fleas may spread typhus and ticks, Lyme disease.

Sexual intercourse is a route of spread for organisms with poor ability to survive outside the body, e.g. *Neisseria gonorrhoeae* or *Treponema pallidum*. Transmission is enhanced by genital ulceration.

Attachment and invasion

Invading microorganisms must attach themselves to host tissues to colonize the body: the distribution of receptors will define the organs which are invaded. *Neisseria gonorrhoeae* adhere to the genital mucosa using fimbriae (non-fimbriate isolates are non-pathogenic). Influenza virus attaches to host cells by its haemagglutinin antigen.

Vibrio cholerae excretes a mucinase to help it reach the enterocyte. *Giardia lamblia* is attached to the jejunal mucosa by a specialized sucking disc. Red cells infected by *Plasmodium falciparum* express a parasite-encoded protein that mediates adherence to host brain capillaries (responsible for cerebral malaria).

Some bacteria form a polysaccharide biofilm that aids colonization of indwelling prosthetic devices, such as catheters.

Motility

The ability to move, to locate new sources of food or in response to chemotactic signals, should enhance pathogenicity: *V. cholerae* is motile by virtue of its flagellum—non-motile mutants are less virulent.

Immune evasion

To survive in the human host, pathogens must overcome the host immune defence. Respiratory bacteria secrete an IgA protease which degrades host immunoglobulin. *Streptococcus pyogenes* expresses protein A which binds host immunoglobulin preventing opsonization and complement activation.

Avoiding destruction by host phagocytes is an important evasive technique. *Streptococcus pneumoniae* has a polysaccharide capsule which inhibits uptake by polymorphonuclear neutrophils (PMNs). Some organisms are specially adapted to survive inside host macrophages, e.g. *Toxoplasma gondii*, *Leishmania donovani*, and *Mycobacterium tuberculosis* escapes into the cytoplasm. The lipopolysaccharide of Gram-negative organisms makes them resistant to the effect of complement. *Trypanosoma* alter surface antigens to evade antibodies.

Damaging the host

Endotoxins stimulate macrophages to produce IL-1 and tumour necrosis factor (TNF), causing fever and shock. Some organisms secrete exotoxins that cause local or distant damage.

4 The laboratory investigation of infection

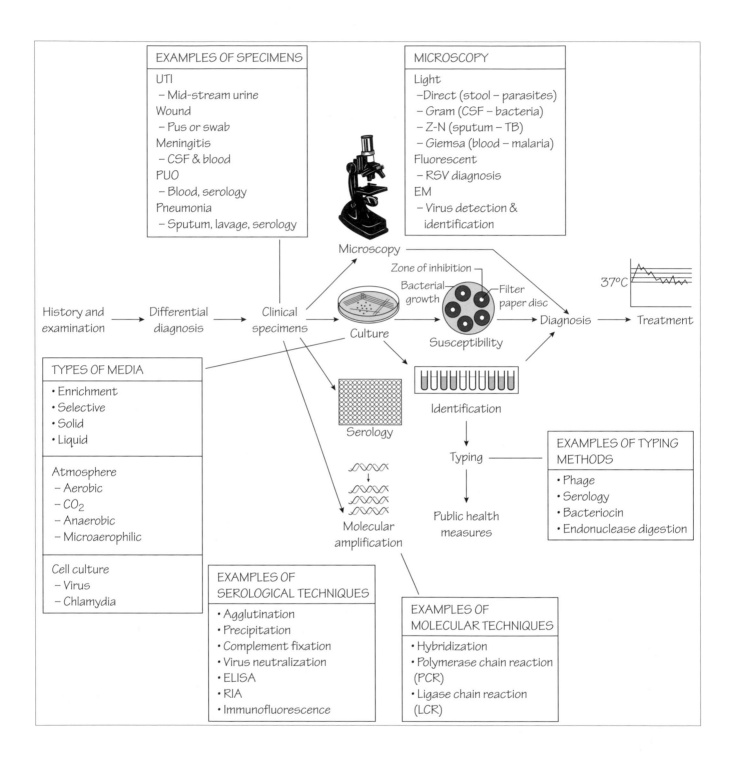

EXAMPLES OF SPECIMENS

UTI
 – Mid-stream urine
Wound
 – Pus or swab
Meningitis
 – CSF & blood
PUO
 – Blood, serology
Pneumonia
 – Sputum, lavage, serology

MICROSCOPY

Light
 –Direct (stool – parasites)
 – Gram (CSF – bacteria)
 – Z-N (sputum – TB)
 – Giemsa (blood – malaria)
Fluorescent
 – RSV diagnosis
EM
 – Virus detection &
   identification

Microscopy

Zone of inhibition
Bacterial growth
Filter paper disc
37°C

History and examination → Differential diagnosis → Clinical specimens → Culture → Susceptibility → Diagnosis → Treatment

TYPES OF MEDIA

• Enrichment
• Selective
• Solid
• Liquid

Atmosphere
 – Aerobic
 – CO$_2$
 – Anaerobic
 – Microaerophilic

Cell culture
 – Virus
 – Chlamydia

Serology

Identification

Typing

Public health measures

Molecular amplification

EXAMPLES OF TYPING METHODS

• Phage
• Serology
• Bacteriocin
• Endonuclease digestion

EXAMPLES OF SEROLOGICAL TECHNIQUES

• Agglutination
• Precipitation
• Complement fixation
• Virus neutralization
• ELISA
• RIA
• Immunofluorescence

EXAMPLES OF MOLECULAR TECHNIQUES

• Hybridization
• Polymerase chain reaction (PCR)
• Ligase chain reaction (LCR)

Specimens

Any tissue or body fluid can be subjected to microbiological investigation. Culture of specimens aims to ensure that any bacteria present will grow as quickly as possible, so hastening identification. This may require the use of enrichment media. In specimens with a normal flora, it is necessary to inhibit the non-pathogens and encourage the growth of pathogens: selective media are required.

Many bacteria do not survive well outside the body: obligate anaerobes may be killed by atmospheric oxygen.

Some organisms are very susceptible to drying (*Neisseria gonorrhoeae*): to protect them during transportation the specimen may be plated onto a suitable medium immediately or inoculated into a transport medium.

Laboratory examination

Specimens may be examined directly, e.g. the presence of adult worms in faeces or of blood in sputum. Microscopic examination is rapid and demands little expensive equipment. It requires considerable technical expertise and is insensitive: it takes a large number of organisms to be present to achieve a positive diagnosis. It also lacks specificity as commensal organisms may be mistaken for pathogens.

Special stains can be used to identify organisms, e.g. Ziehl–Nielsen's method for mycobacteria. Silver methenamine stains the chitin in the cell wall of fungi and *Pneumocystis carinii*. Giemsa is useful for staining malaria and other parasites, such as *Leishmania*.

Immunofluorescence uses antibodies specific to a pathogen that are labelled with a fluorescent marker. The presence of the pathogen is confirmed when examined under ultraviolet light: bound antibody glows as a bright apple-green fluorescence.

Culture

Even when causing severe symptoms, the infecting organism may be present in numbers that are too low to be detected by direct microscopy. Culture amplifies the number of organisms.

Cultivation takes two forms: growth in liquid medium amplifies the number of organisms present; growth on solid media produces individual colonies that can be subcultured for subsequent testing. Most human pathogens are fastidious, requiring media supplemented with peptides, sugars and nucleic acid precursors (present in blood or serum). An appropriate atmosphere must also be provided, e.g. fastidious anaerobes require an oxygen-free atmosphere whereas strict aerobes such as *Bordetella pertussis* require the opposite. Most human pathogens are incubated at 37°C, although some fungal cultures are incubated at 30°C.

Culture allows susceptibility testing (to optimize therapy) and typing (see below).

Identification

The identity of the organism can often predict the clinical course: *Vibrio cholerae* causes a different spectrum of symptoms than *Shigella sonnei*. Identification of certain organisms may lead to public health action; e.g. the isolation of *Neisseria meningitidis* from cerebrospinal fluid.

The process is based on many characteristics including the morphology of colonies isolated on solid media, the Gram stain, the presence of spores and simple biochemical tests, such as catalase, coagulase and oxidase. Precise species identification usually depends on the results of a series of biochemical tests including sugar fermentation tests, enzyme tests, e.g. urease activity, or the detection of bacterial products, e.g. indole.

Susceptibility testing

Organisms are defined as susceptible if a normal dose of an antibiotic is likely to result in cure, moderately resistant if cure is likely with a larger dose and resistant if antibiotic therapy is likely to fail. Paper discs impregnated with antibiotic are placed on agar inoculated with the test organism. The antibiotic diffuses into the surrounding agar and inhibits bacterial growth. The extent of this inhibition reflects the susceptibility of the organism. As clinical response also depends on host factors such *in vitro* tests can only provide an approximate guide.

Serology

Infection can be diagnosed by detecting the immune response to a pathogen. Different methods are used: precipitation, agglutination, complement fixation, virus neutralization, or labelling with enzymes (ELISA), fluorescent markers (see above), or radio-isotopes (radio-immunoassay). The diagnosis is made by detecting a value much higher than the population norms, detecting rising or falling antibody levels in specimens more than a week apart or the presence of specific IgM. Antibodies can be used to detect a specific antigen, e.g. agglutination techniques can be used to detect bacterial capsular antigens in cerebrospinal fluid.

Molecular techniques
Southern blotting

A labelled DNA probe will bind to the specimen if it contains the specific sequence sought. The bound probe is detected by the activity of the label. This is a specific and rapid technique but lacks sensitivity.

Polymerase chain reaction (PCR)

In the PCR test, DNA from the specimen is separated into single strands. Specifically designed primers are added that bind and promote the synthesis of the target DNA. This casade process eventually produces a detectable product. A positive result can be obtained from as little as one copy of the target DNA. Several alternative enzymes have been used including the ligase chain reaction, strand displacement PCR and qβ-replicase.

Typing

It is sometimes necessary to type organisms in order to follow their transmission in the hospital or community (see Chapter 8).

5 Antibacterial therapy

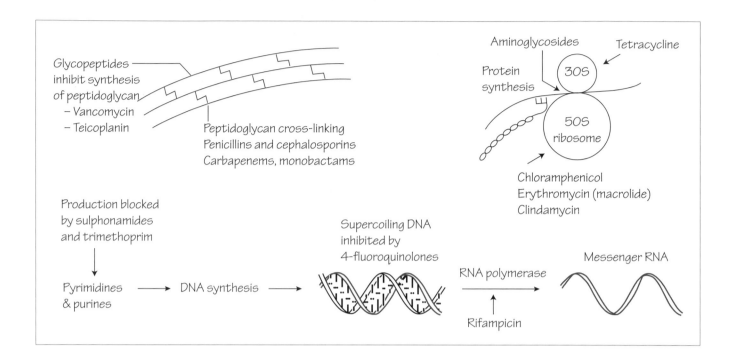

Principles of antibiotic therapy

Antibacterial chemotherapy depends on selective toxicity: the antibiotic interferes with the metabolism of the pathogen but not that of the host. It is best achieved by exploiting bacterial processes that are not present in human cells, e.g. unlike human cells, bacteria possess a cell wall. Inhibiting cell wall synthesis will inhibit the bacterium, but is unlikely to harm the host. Antibiotic treatment is not only effective, but also safe. Most antibiotics have a wide therapeutic index: the dose at which they cause side effects is very much higher than that which inhibits bacterial growth. With the exception of the aminoglycosides, where the serum concentration must be carefully controlled (see below), most antibiotics are remarkably free of serious adverse events.

Mechanisms of action

Tetrahydrofolate is an essential cofactor in the DNA synthesis pathway. As bacteria are unable to utilize host tetrahydrofolate, they are obliged to manufacture it. Antifolate antibiotics, e.g. the sulphonamides and trimethoprim, act by inhibiting the production of tetrahydrofolate so reducing the supply of pyrimidines and purines. They also bind more avidly to bacterial enzymes than to the human version.

To conserve space within the confines of the cell, bacterial DNA is tightly coiled (supercoiled). When transcription occurs the DNA is uncoiled, a process controlled by the tetrameric DNA gyrase enzyme system. 4-Fluoroquinolone antibiotics, e.g. ciprofloxacin, act against this enzyme system. The transcription of messenger RNA is controlled by RNA polymerase: this is inhibited by rifampicin and rifabutin.

Bacterial protein synthesis can be selectively inhibited at the ribosomal level, exploiting the difference between bacterial and mammalian ribosomes. Aminoglycosides act by preventing translation of mRNA into protein. Tetracycline acts by binding to the 30S ribosome thus locking the tRNA to the septal site of mRNA. Chloramphenicol binds to the 50S ribosome preventing attachment of the aminoacyl tRNA and subsequent peptide bond formation. Erythromycin also binds to the 50S ribosome to inhibit protein synthesis by an unknown mechanism.

Bacteria differ from mammals in having a cell wall with peptidoglycan making up the major part of the structure. Penicillins and cephalosporins bind to the enzymes (penicillin-binding proteins) in the peptidoglycan synthesis pathway so inhibiting the process of peptidoglycan cross-linking. Glycopeptides, e.g. vancomycin, interfere with the peptidoglycan cross-linking by a different mechanism.

Metronidazole is only active against anaerobic organisms because it acts as an electron acceptor under anaerobic conditions. It forms toxic metabolites that damage the bacterial DNA.

Adverse events

Mild gastrointestinal upset is probably the most frequent side effect of antibiotic therapy. Rarely, severe allergic reactions may lead to acute anaphylactic shock or serum sickness syndromes.

Gastrointestinal tract

Antibiotic activity can upset the balance of the normal flora within the gut: β-lactams are especially likely to do this, resulting in overgrowth of commensal organisms, such as *Candida* spp. Alternatively, therapy may provoke diarrhoea or, more seriously, pseudomembranous colitis (see Chapter 14).

Skin

Cutaneous manifestations range from mild urticaria or maculopapular, erythematous eruptions to erythema multiforme and the life-threatening Stevens–Johnson syndrome. Most cutaneous reactions are mild and resolve after discontinuation of therapy.

Haemopoietic system

Patients receiving chloramphenicol or antifolate antibiotics may exhibit dose-dependent bone marrow suppression. More seriously, aplastic anaemia may rarely complicate chloramphenicol therapy. High doses of β-lactam antibiotic may induce a granulocytopenia. Antibiotics are a rare cause of haemolytic anaemia.

Renal system

Aminoglycosides may cause renal toxicity by damaging the cells of the proximal convoluted tubule. The elderly, patients with pre-existing renal disease or those who are also receiving other drugs with renal toxicity are at higher risk. Tetracyclines may be toxic to the kidneys.

Liver

Isoniazid and rifampicin may cause a hepatitis: this is more common in patients with pre-existing liver disease. Other agents associated with hepatitis are tetracycline, erythromycin, pyrazinamide, ethionamide and, very rarely, ampicillin. Cholestatic jaundice may follow tetracycline or high-dose fusidic acid therapy.

Choice of therapy

The choice of antibiotic depends on the site of infection, the susceptibilities of the likely infecting organisms, the severity of infection and a history of allergy.

Knowledge of the likely organism that infects a particular site and its antibiotic susceptibility profile usually leads to a rational choice of therapy. An example is the choice of penicillin in the treatment of bacterial pharyngitis, where penicillin-sensitive *Streptococcus pyogenes* is likely to be the infecting organism.

The oral route of administration remains the most commonly used, both in hospital and community practice. Antibiotics may also be given topically for skin infections, *per rectum*, e.g. metronidazole for surgical prophylaxis (see Chapter 8), or vaginally as pessaries. Intravenous therapy may be required in severe infections, such as septicaemia, to ensure adequate antibiotic concentrations. This route may also be chosen for patients unable to tolerate oral therapy, e.g. repeated vomiting. The palatability of paediatric formulations and of patient compliance with frequent or complex regimens must also be considered.

Some sites are difficult to penetrate, such as bone, joints and particularly cerebrospinal fluid. High levels of antibiotic activity are difficult to achieve in abscesses as the blood supply is poor. Low pH may also inhibit antibiotic activity, e.g. aminoglycosides. The problem is magnified when the abscess lies within bone or in the central nervous system.

Many patients consider themselves allergic to one or more antibiotic, most commonly to penicillins. Alternative therapy can usually be selected.

Monitoring therapy

Monitoring of antibiotics may be necessary to ensure that adequate therapeutic levels have been achieved and also to reduce the risk of toxicity. This is especially important where the therapeutic range is close to the toxic range. Serum levels of both aminoglycosides and vancomycin are measured in blood samples taken 1 h before and after intravenous or intramuscular dosage. Peak and trough levels must be adjusted to ensure adequate antibacterial activity and reduce the risk of toxicity, e.g. if the peak is high the dosage may be reduced; a high trough level can be lowered by taking medication less frequently. Levels taken using newer, once-daily regimes are interpreted using normograms.

Serum concentrations are also helpful in the management of partially resistant organisms. If inhibition of an organism only occurs at high antibiotic concentrations, then it follows that it is important to maintain such levels in the circulation. When such an infection arises in a difficult site, e.g. *Pseudomonas* meningitis, antibiotic concentrations may be measured in the cerebrospinal fluid.

Resistance to antibacterial agents

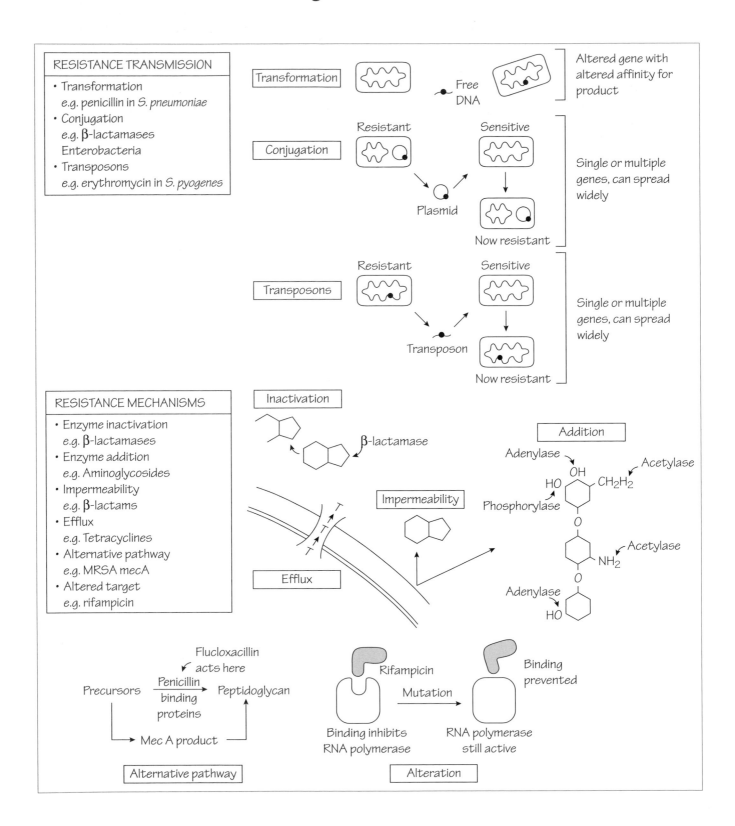

Resistance occurs when a previously susceptible organism is no longer inhibited by an antibiotic. This happens because the bacterial gene pool changes rapidly, facilitated by rapid cell division and the haploid genome. Organisms may transfer genetic material within and between species. Bacteria do not have a deliberate policy to develop 'resistance genes' or 'virulence factors' to advance their species: genetically they play the lottery. Antibiotic use allows the survival and replication of organisms that have accidentally developed mechanisms to avoid destruction.

Transmission of resistance determinants between bacteria

Transformation
Many bacterial species can take up naked DNA and incorporate it into their genome: this is called transformation. It is unlikely that whole 'resistance genes' are taken up in this way. *Streptococcus pneumoniae* takes up part of penicillin-binding protein genes from closely related species. The altered gene produces a penicillin-binding protein which binds penicillin less avidly and so is not inhibited by penicillin to the same extent. The organism is still able to synthesize peptidoglycan and maintain its cell wall in the presence of penicillin. Resistance to penicillin by *Neisseria gonorrhoeae* also develops in the same way.

Conjugation
Plasmids are circular portions of DNA which are found in the cytoplasm. Multiple copies may be present and, following cell division, are found in the cytoplasm of each daughter cell. Many genes are carried on plasmids, including metabolic enzymes, virulence determinants and antibiotic resistance. The process of conjugation occurs when plasmids are passed from one bacterium to another. 'Resistance genes' can spread rapidly this way in populations of bacterial species that share the same environment, e.g. within the intestine. Combined with antibiotic selective pressure, e.g. in hospitals, a multi-resistant population may develop.

Transposons
Transposons are moveable genetic elements able to encode transposition. They can move between the chromosome and plasmids and between bacteria. Many functions, including antibiotic resistance, can be encoded on a transposon. Resistance to methicillin among *S. aureus* and that of *N. gonorrhoeae* to tetracycline probably entered the species by this route. Resistance genes can also be mobilized by bacteriophages.

Mechanisms of resistance
Antibiotic modification
Enzyme inactivation
One of the most common resistance mechanisms occurs when the organism spontaneously produces an enzyme which degrades the antibiotic. Many strains of *S. aureus* produce an extracellular enzyme, β-lactamase, which breaks open the β-lactam ring of penicillin, inactivating it. Many other organisms are capable of expressing enzymes which degrade penicillins and cephalosporins. These include *Escherichia coli*, *Haemophilus influenzae* and *Pseudomonas* spp. The genes that code for these enzymes can be found on mobile genetic elements (transposons) and can be transmitted between organisms of different species.

Enzyme addition
Bacteria may express enzymes that add a chemical group to the antibiotic inhibiting its activity. Bacteria become resistant to aminoglycosides by expressing enzymes that inactivate the antibiotic by adding either an acetyl, an amino or an adenosine group to the aminoglycoside molecule. The different members of the aminoglycoside family differ in their susceptibility to this modification, amikacin being the least susceptible. Aminoglycoside-resistance enzymes are possessed by Gram-positive organisms, such as *S. aureus*, and Gram-negative organisms, such as *Pseudomonas* spp.

Impermeability
Some bacteria are naturally resistant to antibiotics because their cell envelope is impermeable to particular antibiotics. Gram-negative organisms, especially *Pseudomonas* spp., are impermeable to some β-lactam antibiotics. Aminoglycosides enter bacteria by an oxygen-dependent transport mechanism and so have little effect against anaerobic organisms.

Efflux mechanisms
Bacteria, for example *E. coli*, may become resistant to tetracyclines by the acquisition of an inner membrane protein which actively pumps the antibiotic out of the cell.

Alternative pathway
Another common bacterial mechanism is the development of an alternative pathway to circumvent the metabolic block imposed by the antibiotic. *Staphylococcus aureus* becomes resistant to methicillin or flucloxacillin when it acquires the gene *mecA*. This codes an alternative penicillin-binding protein which is not inhibited by methicillin. Although the composition of the cell wall is altered, the organism is still able to multiply. Similar alterations to the penicillin-binding proteins of *Streptococcus pneumoniae* are responsible for resistance in this organism.

Alteration of the target site
Rifampicin acts by inhibiting the β subunit of RNA polymerase. Resistance develops when the RNA polymerase gene is altered by point mutations, insertions or deletions. The new RNA polymerase is not inhibited by rifampicin and resistance occurs.

7 Hospital-acquired infections

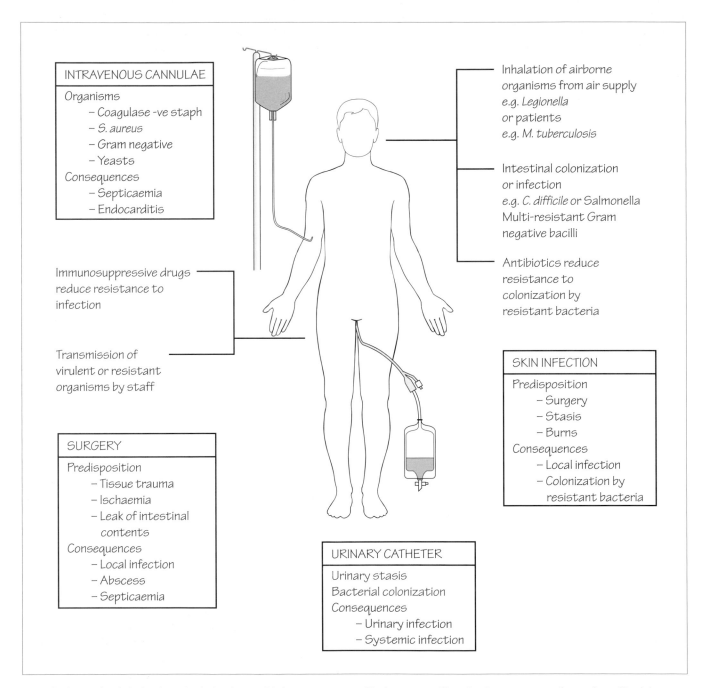

INTRAVENOUS CANNULAE

Organisms
- Coagulase -ve staph
- S. aureus
- Gram negative
- Yeasts

Consequences
- Septicaemia
- Endocarditis

Immunosuppressive drugs reduce resistance to infection

Transmission of virulent or resistant organisms by staff

SURGERY

Predisposition
- Tissue trauma
- Ischaemia
- Leak of intestinal contents

Consequences
- Local infection
- Abscess
- Septicaemia

Inhalation of airborne organisms from air supply e.g. *Legionella* or patients e.g. *M. tuberculosis*

Intestinal colonization or infection e.g. *C. difficile* or Salmonella Multi-resistant Gram negative bacilli

Antibiotics reduce resistance to colonization by resistant bacteria

SKIN INFECTION

Predisposition
- Surgery
- Stasis
- Burns

Consequences
- Local infection
- Colonization by resistant bacteria

URINARY CATHETER

Urinary stasis
Bacterial colonization
Consequences
- Urinary infection
- Systemic infection

Hospital-acquired infection is infection which was not present or incubating at the time of admission. It is very common (up to 25% patients admitted). The most frequent types of infection are urinary tract, respiratory, wound, skin and soft tissue, and septicaemia.

The environment

The potential for person-to-person transmission of organisms within hospitals is enormous (see Chapter 8).

Pathogens will colonize any member of staff, visitor or patient: a medical degree does not confer immunity.

Food supply

Food is usually prepared centrally in the hospital kitchens: patients are at risk of food-borne infection if hygiene standards fall. Antibiotic-resistant organisms can be transmitted by this route.

Air supply

Pathogens, e.g. multidrug resistant TB or respiratory viruses may be transmitted via air. Badly maintained air-conditioning and ventilating systems may also act as a source of pathogens, e.g. *Legionella*.

Fomites

Any inanimate object may become colonized with organisms and act as a vehicle (fomite) for transmission.

Water supply

The water supply in the hospital is a complex system, supplying water to wash-hand basins and showers, central heating and air-conditioning. Additionally, superheated steam at pressure is required for autoclaves. *Legionella* spp. may colonize the system in redundant areas of pipework. Cooling tower systems are a particular source, allowing transmission via the air-conditioning system. To reduce this risk, hot water supplies should be maintained at a temperature above 45°C and cold water supplies below 20°C.

The host

Hospital patients are susceptible to infection as a result of underlying illness or treatment, e.g. patients with leukaemia or taking cytotoxic chemotherapy. Age and immobility may predispose to infection: ischaemia may make tissues more susceptible to bacterial invasion.

Medical activities

Intravenous access

The risk of infection from any intravenous device increases with the length of time it is in position. Having broken the skin's integrity, it provides a route for invasion by skin organisms, e.g. *Staphylococcus aureus*, *S. epidermidis* and *Corynebacterium jeikeium*. Signs of inflammation at the puncture site may be the first evidence of infection. Cannula-related infection can be complicated by septicaemia, endocarditis and metastatic infections, e.g. osteomyelitis. Aseptic technique will reduce the risk of sepsis as will the choice of device, i.e. those without side ports and dead spaces. Maintaining adequate dressing and ensuring good staff hygiene when working with the device is equally important. The state of the cannula site should be regularly inspected: this is particularly important in unconscious patients. Ideally, peripheral lines should be re-sited every 48 h: central and tunnelled lines should be changed when there is evidence of infection.

Urinary catheters

Indwelling urinary catheters provide a route for ascending infection into the bladder. Risks can be minimized by aseptic technique when the catheter is inserted and handled.

Surgery

Surgical patients often have other health problems that are unrelated to their surgical complaint, e.g. asthma or diabetes mellitus, and may predispose them to infection. Surgery is traumatic and carries a risk of infection, e.g. wound infections. In addition, there are the potential complications of the procedure itself, e.g. postoperative ischaemia, that contribute a further risk.

The shorter the preoperative period the lower the risk of acquiring resistant hospital organisms. Elective surgery should be postponed for patients with active infection, e.g. chest infections.

To minimize the risk of infection during an operation, theatres are supplied with a filtered air supply. Staff movement during procedures should be limited to reduce air disturbance. Changing clothing reduces transmission of organisms from the wards. Impervious materials reduce contamination from the skin of the surgical team but are uncomfortable to wear. Some hospitals provide ventilated air-conditioned suits for surgical teams performing prosthetic joint surgery. The length and complexity of the operation influence the risk of infection, as does the skill of the surgeon: the less damage that occurs at the time of operation the lower the risk of infection.

Antibiotic prophylaxis can reduce the risk of postoperative infection. Those chosen should be bactericidal, and penetrate to the required site at sufficient concentrations to be active against organisms normally implicated in infection. There is no evidence that continuing prophylaxis beyond 48 h is beneficial.

'Clean' operations involve only the skin or a normally sterile structure, e.g. a joint. Such operations do not need antibiotic cover unless a prosthetic device is being inserted when antibiotics active against staphylococci should be given.

'Contaminated' operations are those in which a viscus that contains a normal flora is breached. Appropriate antibiotics might be metronidazole with a second-generation cephalosporin for large bowel surgery; cephalosporin alone is satisfactory in upper gastrointestinal tract or biliary tract surgery as anaerobes are rarely implicated.

'Infected operations' are those in which surgery is required to deal with an already infected situation, e.g. drainage of an abscess or repair of a perforated diverticulum. Systemic antibiotics directed against the likely infecting organisms should be prescribed.

Intubation gives organisms access to the lower respiratory system. Postoperative pain, immobility and the effects of anaesthesia may predispose to pneumonia by reducing coughing. Respiratory infections with resistant Gram-negative organisms originating from the hospital environment may also occur.

8 Control of infection in hospital

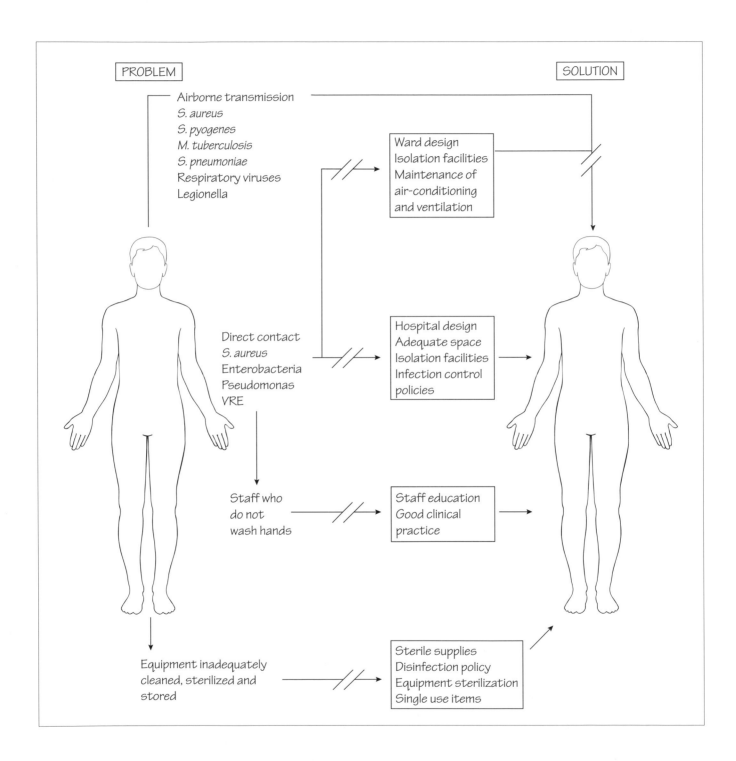

Every hospital should have procedures to ensure that infection is not transmitted within its environment. Together, these form the infection control policy that, if it is to be successful, must have the support of the entire hospital staff. The control of infection team, consisting of a consultant microbiologist or infectious diseases specialist, and specialist nurses, promotes the policy.

The team will arrange enhanced surveillance of particular organisms, e.g. methicillin-resistant *Staphylococcus aureus* (MRSA). It also has a role in hospital planning both physical, e.g. alterations to buildings, and functional, e.g. new clinical services.

Good clinical practice

Infected individuals should be separated from non-infected. Sources, infected or colonized (carriers) must be identified by appropriate screening measures, e.g. routine surveillance specimens from both patients and staff. Infected patients should be isolated (source isolation) and practical measures taken to interrupt possible transmission. Patients who are especially susceptible to infection require protective isolation. Isolation is often difficult to maintain when staff do not adhere to agreed practice. This may be compounded when simple measures, e.g. handwashing, are neglected as a result of work pressures.

Wound and enteric isolation

Patients are nursed in a side room that contains a wash-hand basin and separate toilet facility. Disposable plastic aprons and gloves are used while handling the patient or performing clinical procedures. The gloves and apron are then discarded and hands washed using liquid soap and disposable towels.

Respiratory isolation

In addition to the precautions listed above, hospital staff should also wear a facemask when in the room. If the patient is transferred to another department of the hospital, the patient should wear a facemask. Stricter respiratory isolation methods are necessary to control transmission of multidrug-resistant tuberculosis. This requires the use of negative pressure rooms and effective masks (dust mist masks or personal respirators). Such precautions are especially essential during procedures that are likely to generate aerosols, e.g. bronchoalveolar lavage.

Strict isolation

This form of isolation is designed to prevent the transmission of infections such as viral haemorrhagic fevers. An enclosed isolation unit prevents aerosol transmission of the organism by its enclosed air system and negative pressure, together with strict decontamination procedures.

Protective isolation

Protective isolation is required for patients who are highly susceptible to infection, e.g. neutropenic patients. Protection includes single room isolation, provision of filtered air and measures to control the risk from organisms in food, e.g. resistant Gram-negative organisms in vegetables or *Listeria* in soft cheeses.

Typing

Typing determines if organisms are identical and if cross-infection has occurred. Chosen techniques should be simple to perform, reproducible giving similar results when used in another laboratory.

Simple laboratory typing: using phenotypic markers.

Serological typing: suitable for testing *Shigella flexneri* or the salmonellas.

Phage typing: bacteriophages lyse the bacteria they infect: this phenomenon is used in phage typing, e.g. for staphylococci and some species of *Salmonella*.

Colicine typing: some bacteria produce protein antibiotics, e.g. colicines that inhibit closely related bacteria. This can be used to type *Shigella*.

Molecular typing: restriction endonuclease enzymes are used to digest genomic, plasmid DNA, or ribosomal RNA giving a characteristic pattern. Identical organisms will have identical band patterns.

Sterilization and disinfection

Sterilization

Sterilization inactivates all infectious organisms and is achieved by autoclaving or irradiation. In the autoclave items, e.g. surgical instruments, are heated with superheated pressurized steam to inactivate any contaminating infectious material. Delicate instruments can be sterilized at low pressures and temperatures in specialized autoclaves that deliver steam together with formaldehyde. Perishable materials, such as plastic cannulae, syringes or prosthetic devices, are sterilized using γ-irradiation during commercial manufacture. Aldehydes, e.g. glutaraldehyde and formaldehyde, are capable of sterilizing instruments if they are adequately cleaned first and the equipment is immersed for a sufficient length of time.

Disinfection

This is the process of reducing the number of infectious particles. Simple washing with soaps or detergents is the most important component in disinfection. Disinfectants are chemicals that kill or inhibit microbes. They are used where it would be impossible to achieve sterile conditions, e.g. skin preparation before surgery, or after spillage of biological fluid (urine, blood or faeces) over an inanimate surface. Hypochlorite compounds (sodium hypochlorite, bleach) are most active against viruses, are also useful after spillage, but are corrosive to metals. Halogen compounds, such as iodine, are active against bacteria, including spore-bearing organisms, but are relatively slow acting. They are used in disinfection of skin. Phenolic disinfectants are highly active against bacteria and are used to disinfect contaminated surfaces in the hospital and in bacteriology laboratories. Alcohol (70%) acts rapidly against bacteria and viruses and is useful in disinfecting skin preoperatively. Chlorhexidine is active against bacteria, especially staphylococci: it is also used for disinfection of the skin.

9 Control of infection in the community

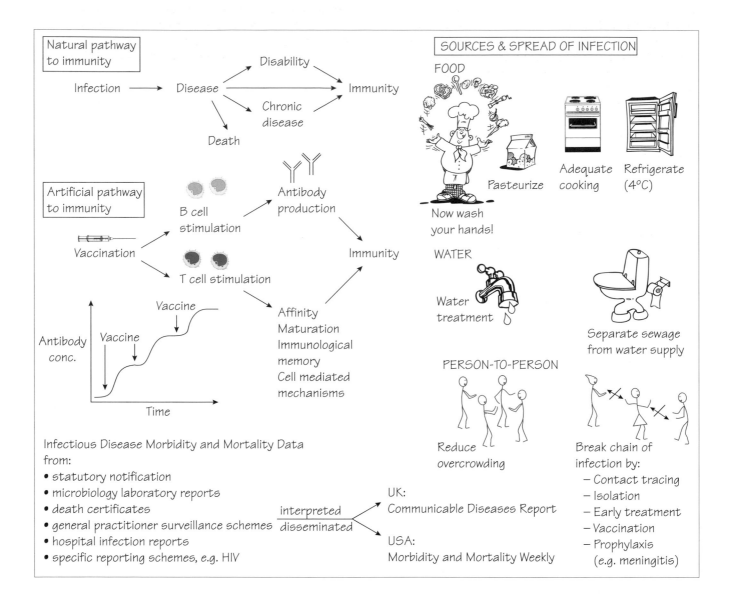

Social and environmental factors

Important factors in reducing the burden of infectious disease are improvements in social and environmental conditions, e.g. improved sanitation reduces the risk of diarrhoeal diseases and better housing reduces the spread of tuberculosis. Better nutrition means that the population is less susceptible to disease.

Paradoxically, the morbidity from some infectious diseases may rise as living conditions improve. This occurs when the complication rate is greater in adults than in children, e.g. paralytic poliomyelitis (see Chapter 29).

Health education

There are many effective infection-related health educa-tion programmes covering safe sex, needle exchange, advice to pregnant women, guidance on food hygiene and advice to travellers.

Food safety

Food safety legislation has been harmonized across the European Community (the Food Safety Act in the UK). The law is enforced in food premises by environmental health officers (EHOs) and officials of the Ministry of Agriculture, Fisheries and Food (MAFF) on farms. Milk pasteurization reduces the risk of infection with *Mycobacterium bovis* and *Campylobacter* spp.

Vector control

This is important where arthropods transmit infections. Travellers to the tropics can reduce the risk of infection by taking measures to avoid insect bites. Attempts to control insect populations using pesticides have usually been unsuccessful because of resistance.

Immunization

Many infectious diseases are controlled by vaccination, e.g. polio and diphtheria. Smallpox has been completely eradicated by this means. Immunization may be achieved *passively* by administration of an immunoglobulin preparation, or *actively* by vaccination.

Immunoglobulins

Immunoglobulins provide short-term protection against certain infections and are useful in the management of immune disorders. Human immunoglobulin (HIG) is prepared from pooled plasma and contains antibodies to viruses that are prevalent in the general population. Specific immunoglobulins, prepared from hyperimmune donors, are available for postexposure management, e.g. hepatitis B, varicella-zoster and tetanus. Tick-borne encephalitis immunoglobulin is available in countries where the disease is endemic.

Vaccination

The purpose of vaccination is to produce immunity in subjects without the complications of natural infection. Vaccines are derived from whole viruses and bacteria, or their antigenic components (acellular).

Live vaccines consist of strains with little pathogenicity (attenuated), e.g. mumps, measles, rubella, polio, yellow fever. They usually produce durable immunity after a single dose. Live vaccines may cause disease in immunocompromised patients and are avoided in pregnancy because of the risk of fetal infection.

Non-replicating vaccines contain either inactivated whole organisms (pertussis) or antigenic components (capsular polysaccharide of *Streptococcus pneumoniae*). The toxins of tetanus and diphtheria are inactivated to produce toxoids that do not cause symptoms but are fully immunogenic. The immunogenicity of acellular vaccines can be increased by conjugation with proteins (*Haemophilus influenzae*). Genetic engineering is used for acellular vaccine production (hepatitis B). These vaccines are safe in immunocompromised patients because they are unable to replicate. Multiple doses may be required for optimum immunogenicity.

The aim of an immunization programme may be eradication, elimination or containment. Eradication is total absence of the organism in humans, animals and the environment, e.g. smallpox. In elimination, the disease has disappeared but the organism remains, e.g. animal hosts, the environment. Universal immunization is adopted for most childhood infections and selective programmes for those at risk from disease, e.g. hepatitis B in health-care workers. National immunization programmes are updated regularly.

Chemoprophylaxis

This is used for control of more serious infections, e.g. diphtheria, meningococcal disease. It aims to eliminate carriage of pathogens, e.g. rifampicin for meningococcal contacts.

Outbreak investigation

Basic epidemiological information is collected, e.g. onset of symptoms, age/sex, place of residence, detailed food history.

A hypothesis of causation is tested by a case–control or cohort study. In a case–control study, exposure histories are sought from cases and healthy controls. The relative risk of exposure is calculated for each group. Case–control studies are suited to investigation of uncommon infections, such as botulism. Cohort studies compare the disease outcome between those exposed and not exposed. They are often used to investigate outbreaks with a high attack rate, e.g. food poisoning incidents.

The role of national agencies

Most countries have a national system to control communicable diseases. It has four main functions:
- surveillance of communicable diseases
- investigation of outbreaks
- surveillance of immunization programmes
- epidemiology research and training

Close collaboration between food and agriculture control agencies and the human infection control agency is required for zoonotic infections.

10 Staphylococcus

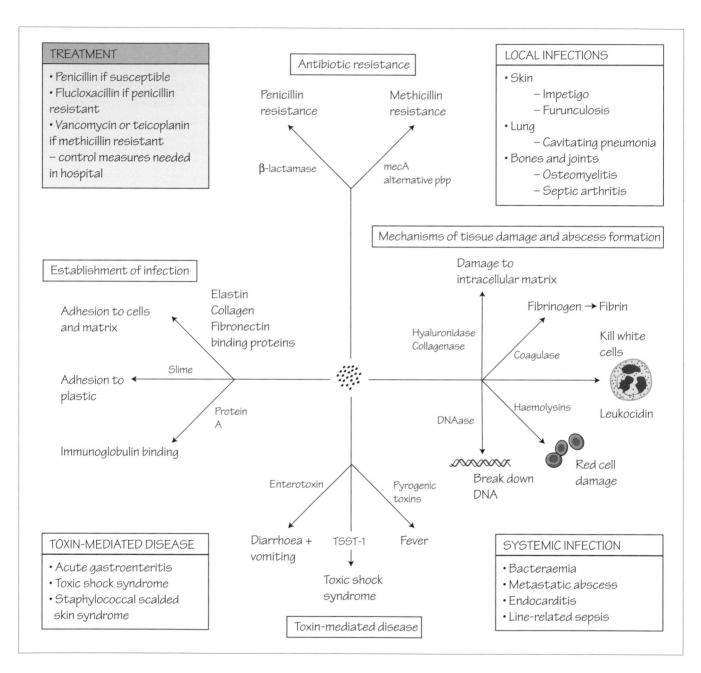

TREATMENT
- Penicillin if susceptible
- Flucloxacillin if penicillin resistant
- Vancomycin or teicoplanin if methicillin resistant
  - control measures needed in hospital

Antibiotic resistance

Penicillin resistance — β-lactamase

Methicillin resistance — mecA alternative pbp

LOCAL INFECTIONS
- Skin
  - Impetigo
  - Furunculosis
- Lung
  - Cavitating pneumonia
- Bones and joints
  - Osteomyelitis
  - Septic arthritis

Mechanisms of tissue damage and abscess formation

Damage to intracellular matrix

Hyaluronidase Collagenase

Fibrinogen → Fibrin — Coagulase

Kill white cells — Leukocidin

Haemolysins — Red cell damage

DNAase — Break down DNA

Establishment of infection

Adhesion to cells and matrix — Elastin Collagen Fibronectin binding proteins

Adhesion to plastic — Slime

Immunoglobulin binding — Protein A

TOXIN-MEDIATED DISEASE
- Acute gastroenteritis
- Toxic shock syndrome
- Staphylococcal scalded skin syndrome

Enterotoxin — Diarrhoea + vomiting

TSST-1 — Toxic shock syndrome

Pyrogenic toxins — Fever

Toxin-mediated disease

SYSTEMIC INFECTION
- Bacteraemia
- Metastatic abscess
- Endocarditis
- Line-related sepsis

Microscopically, these non-sporing, non-motile, Gram-positive cocci form clusters. They form part of the normal skin flora of humans and animals.

Classification
Staphylococci are part of the family Micrococcaceae. There are more than 26 species but only a few are associated with human disease. *Staphylococcus aureus* is the most invasive species and is differentiated from other species by possessing the enzyme coagulase.

Staphylococcus aureus
This species was once thought to be the only pathogen in the genus. Asymptomatic carriage of *S. aureus* is common and found in up to 40% of normal people in the nose, on the skin, in the axilla or perineum.

Pathogenesis
Staphylococcus aureus produces coagulase which catalyses the conversion of fibrinogen to fibrin and may help the organism to form a protective barricade. It also has recep-

tors for the host cell surface proteins, e.g. fibronectin that help the organism adhere. It produces extracellular lytic enzymes, which break down host tissues and aid invasion. Some strains produce potent exotoxins, which may cause a toxic shock syndrome. Enterotoxins may also be produced, causing diarrhoea.

Clinical importance
Staphylococcus aureus causes a wide range of infectious syndromes. Skin infections are favoured by warm moist conditions or when the integrity of the skin is broken by disease, e.g. eczema, by surgical wounds or by intravenous devices. Impetigo may occur in healthy skin: infection is transmitted from person to person. *Staphylococcus aureus* pneumonia is rare but may follow influenza. It progresses rapidly with cavity formation and a high mortality. *Staphylococcus aureus* endocarditis is equally rapid and destructive and may follow illicit intravenous drug-use or infections from intravenous devices. *Staphylococcus aureus* is the most common agent of osteomyelitis and septic arthritis (see Chapter 45).

Laboratory diagnosis
Staphylococcus aureus grows readily on most laboratory media; as it is tolerant of high salt concentrations, media can be made selective in this way. Most *S. aureus* ferment mannitol: incorporation of mannitol and an indicator dye will enable them to be selected for subculture. Organisms are identified by possession of coagulase, DNAase, and catalase enzymes, typical 'cluster of grapes' morphology on Gram stain, and biochemical testing. *Staphylococcus aureus* can be typed by using the lytic properties of an international battery of phages or DNA restriction profiles.

Antibiotic susceptibility
The history of susceptibility of *S. aureus* is a lesson in the history of antimicrobial chemotherapy.
1 Initially susceptible to penicillin, but β-lactamase producing strains soon predominated.
2 Introduction of methicillin and related agents, e.g. flucloxacillin, replaced penicillin as the drug of choice. This is still the drug of choice in sensitive strains.
3 Methicillin-resistant *S. aureus* (MRSA) emerged. Resistance is caused by possession of the *mecA* gene which codes for a low-affinity penicillin-binding protein. Some MRSA have epidemic (EMRSA) potential. Vancomycin or teicoplanin may be required for these strains.
4 Glycopeptide resistance may be emerging with recent reports of reduced susceptibility to vancomycin.
Other antibiotics that are effective include aminoglycosides, erythromycin, clindamycin, fusidic acid, chloramphenicol, and tetracycline.

In methicillin-sensitive strains, first and second generation cephalosporins are effective. Fusidic acid may be given with another agent in bone and joint infections. Treatment should be guided by sensitivity testing.

Prevention and control
Staphylococcus aureus spreads by airborne transmission and via the hands of health-care workers. Patients colonized or infected with MRSA should be isolated in a side-room with wound and enteric precautions. Staff may become carriers and disseminate the organism widely in the hospital environment. Carriage may be eradicated by using topical mupirocin and chlorhexidine.

Staphylococcus epidermidis
Staphylococcus epidermidis is the most important of the coagulase-negative staphylococci. Once dismissed as contaminants, they are now recognized as pathogens if conditions favour their multiplication.

Clinical importance
S. epidermidis causes infection of intravenous cannulae, long-standing intravascular prosthetic devices, ventriculo-peritoneal shunts and prosthetic joints. This may lead to bacteraemia and require the removal of the prosthesis.

Laboratory diagnosis
Staphylococcus epidermidis grows readily on laboratory media; coagulase is not produced. Speciation is by biochemical testing. DNA restriction patterns or other molecular techniques may be needed to determine whether strains are identical. *S. epidermidis* is a common contaminant and thus careful evaluation of its clinical relevance is important.

Antibiotic susceptibility
This group of organisms are uniformly susceptible to vancomycin and usually to teicoplanin; it can be susceptible to any of the agents used for *S. aureus* infection, but this is unpredictable. Treatment must be guided by *in vitro* testing.

Staphylococcus haemolyticus
Less common than *S. epidermidis*, *S. haemolyticus* causes a similar disease pattern. It differs from *S. epidermidis* in that it causes haemolysis on blood agar. More importantly, it is naturally resistant to teicoplanin: significant infections require vancomycin therapy.

Staphylococcus saprophyticus
This coagulase-negative staphylococcus is a common cause of urinary tract infection in young women. It is distinguished by resistance to novobiocin.

11 Streptococcal infections

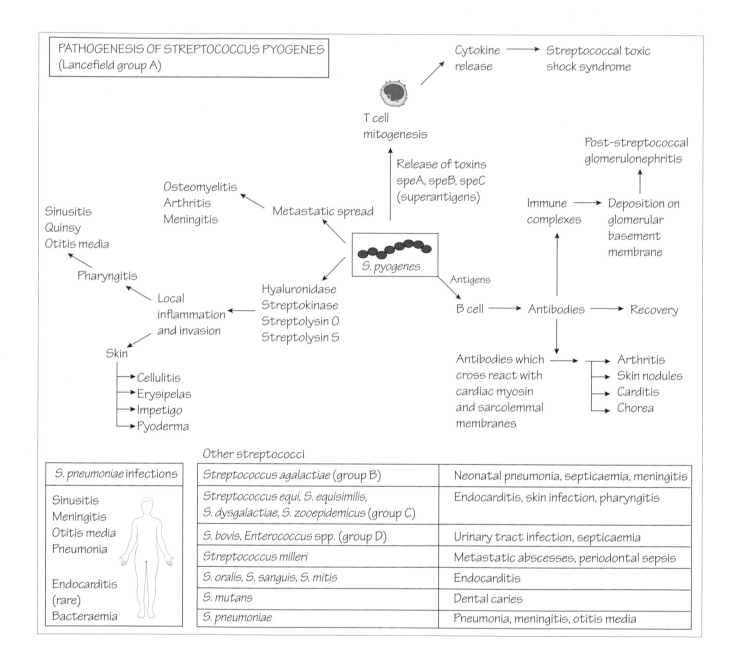

PATHOGENESIS OF STREPTOCOCCUS PYOGENES
(Lancefield group A)

Cytokine release → Streptococcal toxic shock syndrome

T cell mitogenesis

Release of toxins speA, speB, speC (superantigens)

Post-streptococcal glomerulonephritis

Immune complexes → Deposition on glomerular basement membrane

Osteomyelitis
Arthritis
Meningitis

Metastatic spread

S. pyogenes

Sinusitis
Quinsy
Otitis media

Pharyngitis

Local inflammation and invasion

Hyaluronidase
Streptokinase
Streptolysin O
Streptolysin S

Antigens

B cell → Antibodies → Recovery

Antibodies which cross react with cardiac myosin and sarcolemmal membranes → Arthritis / Skin nodules / Carditis / Chorea

Skin
→ Cellulitis
→ Erysipelas
→ Impetigo
→ Pyoderma

Other streptococci

S. pneumoniae infections	Streptococcus agalactiae (group B)	Neonatal pneumonia, septicaemia, meningitis
Sinusitis	Streptococcus equi, S. equisimilis, S. dysgalactiae, S. zooepidemicus (group C)	Endocarditis, skin infection, pharyngitis
Meningitis		
Otitis media	S. bovis, Enterococcus spp. (group D)	Urinary tract infection, septicaemia
Pneumonia	Streptococcus milleri	Metastatic abscesses, periodontal sepsis
	S. oralis, S. sanguis, S. mitis	Endocarditis
Endocarditis (rare)	S. mutans	Dental caries
Bacteraemia	S. pneumoniae	Pneumonia, meningitis, otitis media

These Gram-positive cocci are arranged in pairs and chains. Although facultative anaerobes, they are fastidious, requiring rich blood-containing media. Swabs from the site of infection (throat/wound, etc.) and blood culture should be taken. Colonies are distinguished by the type of haemolysis; complete (β) or incomplete type (α). Biochemical and serological (Lancefield grouping) tests are used for further identification.

Streptococcus pyogenes

Carried asymptomatically in the pharynx in 5–30% of the population, *S. pyogenes* is transmitted by the aerosol route and by contact. Infection is most common in children but can arise at any age.

Pathogenesis

Streptococcus pyogenes carries a group A carbohydrate antigen (Lancefield's antigen) and is surrounded by the M protein antigen, which prevents leucocyte phagocytosis. Antibodies to particular M proteins are protective against further infection with the same type. Several toxins may be produced: erythrogenic toxin, streptococcal pyrogenic exotoxins A, B and C.

Clinical presentation

S. pyogenes is associated with three types of disease:

1 Infection: it is the most common bacterial cause of pharyngitis. It also causes erysipelas, impetigo, cellulitis, wound infections and rarely, necrotizing fasciitis. Septicaemia may occur and result in metastatic infections, e.g. osteomyelitis.

2 Toxin-mediated disease in association with infection: erythrogenic toxin causes scarlet fever; pyrogenic toxin-producing strains are associated with streptococcal shock.

3 Postinfectious immune-mediated disease: rheumatic fever, glomerulonephritis or erythema nodosum are thought to be immune mediated because antibodies cross-react with host tissues.

Prevention and control

Streptococcus pyogenes can spread rapidly in surgical and obstetric wards; infected or colonized patients should be isolated in a side-room until 48 h after initiation of antibiotic therapy. Prompt treatment prevents secondary immune disease, e.g. rheumatic fever. Benzyl penicillin is the treatment of choice. Amoxicillin may be used for oral therapy in less severe infections.

Streptococcus pneumoniae

Humans are the only host of *S. pneumoniae*; most carriage is asymptomatic. Children under 1 year of age are especially susceptible to acute pneumonia. Complement deficiency, agammaglobulinaemia, HIV and splenectomy predispose to severe infection. The capsule is the main pathogenicity determinant and there are more than 90 different types with varying invasive potential. Toxins, e.g. pneumolysin, neuraminidase, hyaluronidase and adhesins (e.g. pneumococcal surface protein A) are important in the pathogenesis of disease.

Clinical features

Acute otitis media, sinusitis and acute pneumonia are the most common infections. Direct or haematogenous spread can give rise to meningitis. Bacteraemia is an important complication with a high mortality despite treatment (see Chapter 40).

Antibiotic susceptibility and treatment

Once universally susceptible to penicillin, significant numbers have developed a resistant penicillin-binding protein gene (see Chapter 6). Also susceptible to erythromycin, cephalosporins, tetracycline, rifampicin and chloramphenicol, but multiple drug resistance is growing. Penicillin is still the treatment of choice; cephalosporins are used for meningitis caused by resistant strains.

Prevention and control

A polyvalent capsular polysaccharide vaccine is effective in adults but less so in immunocompromised patients and children under 2 years. A conjugate vaccine currently under trial may be more effective.

Streptococcus agalactiae

Streptococcus agalactiae (group B streptococcus) is a normal gut commensal and may be found in the female genital tract. Early (up to 1 week) perinatal infection causes pneumonia or septicaemia associated with high mortality; later infections cause meningitis. The polysaccharide antiphagocytic capsule is the main pathogenicity determinant. Babies of mothers with antibody to the four capsular types (Ia, Ib, II and III) are protected from infection.

Clinical features and diagnosis

Infected neonates may initially lack the classical clinical signs of sepsis, e.g. fever and the bulging fontanelle of meningitis. A chest X-ray may demonstrate pneumonia and specimens of blood, cerebrospinal fluid, amniotic fluid and gastric aspirate should be cultured. Antigen detection tests are available and can be applied to body fluids for rapid diagnosis.

Treatment and prevention

Neonatal group B streptococcal sepsis requires empirical therapy including a penicillin and aminoglycoside. Perinatal penicillin can prevent invasive infection but should be targeted at high-risk babies.

Oral streptococci

Metastatic abscesses

The *Streptococcus milleri* group of organisms colonize the mouth. They may spread systemically causing brain, lung or liver abscesses.

Enterococcus spp.

Enterococci possess a group D carbohydrate cell wall antigen.

Of more than 12 species, *E. faecalis* and *E. faecium* are the main pathogens causing urinary tract infection, wound infection and endocarditis. Enterococci are emerging as hospital pathogens with some species (*E. faecium*) resistant to commonly used antibiotics.

12 *Corynebacterium*, *Listeria* and *Bacillus*

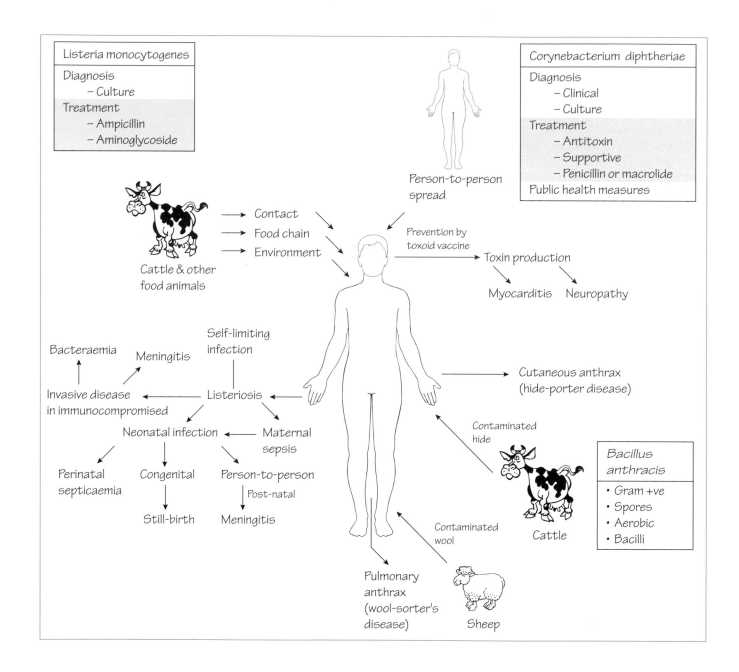

Listeria monocytogenes
Diagnosis
– Culture
Treatment
– Ampicillin
– Aminoglycoside

Corynebacterium diphtheriae
Diagnosis
– Clinical
– Culture
Treatment
– Antitoxin
– Supportive
– Penicillin or macrolide
Public health measures

Person-to-person spread

Cattle & other food animals

Contact
Food chain
Environment

Prevention by toxoid vaccine

Toxin production

Myocarditis   Neuropathy

Bacteraemia   Meningitis

Self-limiting infection

Invasive disease in immunocompromised

Listeriosis

Cutaneous anthrax (hide-porter disease)

Contaminated hide

Neonatal infection   Maternal sepsis

Perinatal septicaemia   Congenital   Person-to-person

Post-natal

Still-birth   Meningitis

Bacillus anthracis
• Gram +ve
• Spores
• Aerobic
• Bacilli

Cattle

Contaminated wool

Pulmonary anthrax (wool-sorter's disease)

Sheep

Corynebacterium spp.

There are many species in this genus. All are non-sporing, non-motile and non-capsulate Gram-positive pleomorphic bacilli, arranged in an irregular pattern.

Corynebacterium diphtheriae

Corynebacterium diphtheriae is transmitted by the respiratory route or following direct contact from cutaneous lesions.

Pathogenesis
Diphtheria is caused by a strain of *C. diphtheriae* containing a bacteriophage encoding diphtheria toxin.

Clinical features and management
Infection may occur on the skin, nasopharynx or larynx. In the throat, it causes acute inflammation and necrosis shown by a green, black 'pseudomembrane' on the posterior wall of the pharynx. The membrane can cause respiratory obstruction. Diphtheria produces a toxaemia: clinical

severity is directly related to toxin production. The toxin, which affects protein synthesis, has a direct action on the myocardium causing myocarditis, and on the peripheral nervous system causing neuropathy and paralysis. Cutaneous infection is often asymptomatic. Management is based on isolation and giving antitoxin and erythromycin. Intensive care support may be required.

Laboratory diagnosis
C. diphtheriae, which produces black colonies on media containing tellurite (e.g. Hoyle's), is identified by biochemical tests. Toxin production is confirmed by agar immuno-diffusion (Elek's test) or detection of toxin gene by PCR. Löffler's serum media may be useful in early isolation.

Prevention and control
Diphtheria is prevented by vaccination with a toxoid as in the childhood immunization schedule. Immunity is long-lasting but adults may require a booster. Contacts of cases must be isolated and screened.

Corynebacterium jeikeium
Naturally resistant to most antibiotics except vancomycin, this organism causes prosthetic infection and bacteraemia in immunocompromised individuals.

Other corynebacteria
Rarely *C. ulcerans* may cause diphtherial pharyngitis: likewise *C. pseudotuberculosis* may cause suppurative granulomatous lymphadenitis. *Rhodococcus equi* has been associated with a severe cavitating pneumonia in AIDS patients.

Listeria
A Gram-positive, non-sporing, motile, facultative anaerobic bacillus which can grow at low temperatures (4–10°C). *Listeria monocytogenes* is associated with human disease.

Epidemiology
Listeria spp. are found in soil or in foodstuffs where contamination by animal faeces has occurred. Cross contamination of food products may occur. Infection follows consumption of contaminated food, e.g. soft cheeses.

Clinical features
L. monocytogenes causes a mild, self-limiting, infectious mononucleosis-like syndrome. Rarely, acute pyogenic meningitis or encephalitis (with a high mortality) develops in patients with reduced cell-mediated immunity. Bacteraemia in pregnancy is associated with intra-uterine death, premature labour and neonatal infection similar to group B streptococci (see Chapter 38).

Laboratory diagnosis
Listeria grow readily on simple laboratory media, exhibiting a narrow zone of haemolysis on blood agar. It can be selected by incubating at low temperature but selective media allows faster isolation. Further identification is made by biochemical testing. Serological typing uses antibodies to the somatic O antigens.

Management
Listeria spp. are susceptible to ampicillin and gentamicin but resistant to the cephalosporins, penicillin and chloramphenicol. Patients with symptoms of meningitis, in whom listeriosis is a possible diagnosis, should have ampicillin incorporated into their drug regimen. Laboratory confirmed listeriosis should be treated with ampicillin and gentamicin for 2–6 weeks depending on the site of infection.

Prevention and control
Listeriosis can be prevented by good food hygiene, effective refrigeration and adequate re-heating of pre-prepared food. Individuals who are at particular risk, such as pregnant women and the immunocompromised, should avoid high-risk foods.

Bacillus
These Gram-positive, aerobic bacilli are able to survive in adverse environmental conditions by forming spores.

Bacillus anthracis
Bacillus anthracis is a soil organism that, under certain climatic conditions, multiplies to cause anthrax in herbivores. Humans become infected from contaminated animal products. Pathogenicity depends on three bacterial antigens: the 'protective antigen' and oedema factor (both toxins) and the antiphagocytic poly-D-glutamic acid capsule. Inoculation of *B. anthracis* into minor skin abrasions produces a necrotic, oedematous ulcer with regional lymphadenopathy (hide-porter's disease). Inhalation of anthrax spores develops into fulminant pneumonia and septicaemia (wool-sorter's disease). The diagnosis must be made in a laboratory equipped and specialized in handling this organism. Treatment is with penicillin, erythromycin or tetracycline. Anthrax is prevented by animal vaccination treatment of animal products and vaccination of humans at high risk.

Bacillus cereus
Bacillus cereus causes a heat stable toxin. Typically, it multiplies in par-boiled rice (during preparation of fried rice), causing a self-limiting food poisoning. Vomiting occurs 6h after exposure, followed by diarrhoea (18h).

13 Mycobacteria

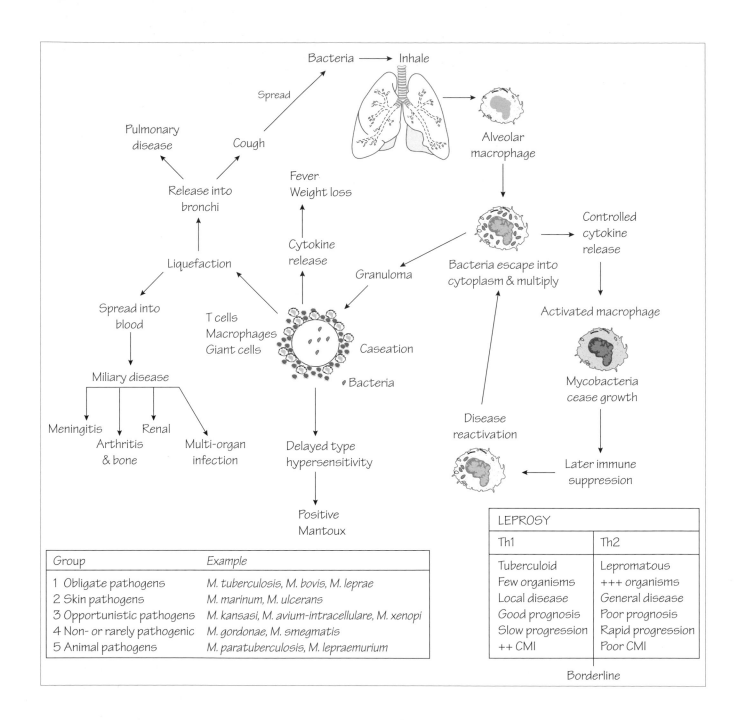

Group	Example
1 Obligate pathogens	M. tuberculosis, M. bovis, M. leprae
2 Skin pathogens	M. marinum, M. ulcerans
3 Opportunistic pathogens	M. kansasi, M. avium-intracellulare, M. xenopi
4 Non- or rarely pathogenic	M. gordonae, M. smegmatis
5 Animal pathogens	M. paratuberculosis, M. lepraemurium

LEPROSY	
Th1	Th2
Tuberculoid	Lepromatous
Few organisms	+++ organisms
Local disease	General disease
Good prognosis	Poor prognosis
Slow progression	Rapid progression
++ CMI	Poor CMI

Borderline

Mycobacteria possess a lipid-rich cell wall that retains some dyes, even resisting decolorization with acid (acid-fast). There are more than 50 species: most are environmental organisms that rarely cause human infection.

Mycobacterium tuberculosis
Epidemiology and pathogenesis
Tuberculosis is spread from person to person by the aerosol route. The lung is the first site of infection: most infections resolve with local scarring (the primary complex). Infection may disseminate from the primary focus throughout the body (miliary spread): this may resolve spontaneously or develop into localized infection, e.g. meningitis. Resistance to tuberculosis depends on T cell function: disease may reactivate if immunity falls (estimated 10% lifetime risk of reactivation). Immunocompromised individuals, e.g.

HIV positive, are more likely to develop symptomatic disease.

Mycobacterium tuberculosis is ingested by macrophages but escapes from the phagolysosome to multiply in the cytoplasm. The intense immune response causes local tissue destruction (cavitation in the lung) and cytokine-mediated systemic effects (fever and weight loss). Many antigens have been identified as possible virulence determinants, e.g. lipoarabinomannan (stimulates cytokines); superoxide dismutase (promotes intramacrophage survival).

Clinical features

Mycobacterium tuberculosis may affect every organ of the body: it mimics both inflammatory and malignant diseases. Pulmonary tuberculosis may present with a chronic cough, haemoptysis, fever and weight loss or as recurrent bacterial pneumonia. Untreated, the infection follows a chronic, deteriorating course. Tuberculous meningitis presents with fever and slowly deteriorating level of consciousness. Kidney infection may lead to signs of local infection, fever and weight loss, complicated by ureteric fibrosis and hydronephrosis. The lumbosacral spine is a common site of bone infection: progression may cause vertebral collapse and nerve compression. Additionally, pus may spread under the psoas sheath to appear as a groin swelling (psoas abscess). Infection of large joints may lead to a destructive arthritis. In abdominal infection, mesenteric lymphadenopathy and chronic peritonitis may present as fever, weight loss, ascites and intestinal malabsorption. Disseminated infection (miliary disease) can occur without evidence of active lung infection.

Laboratory diagnosis

Specimens are stained by Ziehl–Nielsen's method, then cultured on lipid-rich (egg-containing) medium with malachite green (Löwenstein–Jensen) to suppress other organisms. Growth can be detected more quickly in broth culture by radiometric or fluorescence methods. Susceptibility is tested on slopes of L–J medium or in radiometric broth. Polymerase chain reaction and Southern hybridization are helpful in the rapid diagnosis of mycobacterial infection. New molecular methods for rapid susceptibility testing are now available. *M. tuberculosis* can be typed by an internationally standardized restriction fragment length polymorphism (RFLP) method.

Treatment and prevention

The standard regimen for pulmonary infection is rifampicin and isoniazid for 6 months with ethambutol and pyrazinamide for the first 2 months. Regimes for other sites are similar, taking into account drug penetration, e.g. into CSF. Treatment for multidrug-resistant tuberculosis (MDRTB) is guided by susceptibility tests or the pattern of an an epidemic strain.

Vaccination with attenuated bacille–Calmette–Guérin (BCG) strain may protect against miliary spread but trials in some countries have shown no benefit. Patients at high risk of developing tuberculosis may be given prophylaxis with isoniazid and rifampicin. HIV patients may benefit from long-term prophylaxis with rifabutin or clarithromycin.

Mycobacterium leprae

Mycobacterium leprae cannot be cultivated in artificial medium. The organism attacks peripheral nerves causing anaesthesia. Digital destruction and deformity follow, leaving the patient severely disabled. The end result depends on the immune response, forming a spectrum from 'tuberculoid' dominated by a Th1 response, through 'borderline' to 'lepromatous' dominated by a Th2 response. Patients with lepromatous disease have poor cell mediated immunity, no granuloma, and generalized disease (the leonine facies, depigmentation and anaesthesia). Diagnosis is by Ziehl–Nielsen's stain of a split-skin smear and histological examination of skin biopsy. Treatment with rifampicin, dapsone and clofazimine render the patient non-infectious rapidly but cannot alter nerve damage and deformity, which must by managed by remedial surgery.

Mycobacterium avium-intracellulare complex (MAIC)

This includes *Mycobacterium avium*, *M. intracellulare* and *M. scrofulaceum*. Some are natural pathogens of birds, others are environmental saprophytes. A common cause of mycobacterial lymphadenitis in children, they cause osteomyelitis in immunocompromised patients and chronic pulmonary infection in the elderly. In advanced HIV disease, they cause disseminated infection and bacteraemia. MAIC are naturally resistant to many antituberculosis agents: multi-drug regimens should be used including rifabutin, clarithromycin and ethambutol. Lymphadenitis may require surgery.

Mycobacterium kansasi, M. malmoense, M. xenopi

These species cause an indolent pulmonary infection resembling tuberculosis in individuals predisposed by chronic lung disease, e.g. bronchiectasis, silicosis and obstructive airway disease. Initial therapy with standard drugs may have to be adjusted following susceptibility tests.

Mycobacterium marinum, M. ulcerans

M. marium causes a chronic granulomatous infection of the skin acquired from rivers, poorly maintained swimming pools or fish tanks: it is characterized by encrusted pustular lesions. *M. ulcerans* infection is associated with farming in Africa and Australia. The lower limb is usually affected with a papular lesion, which ulcerates and may destroy underlying tissue including bone.

14 *Clostridium* species

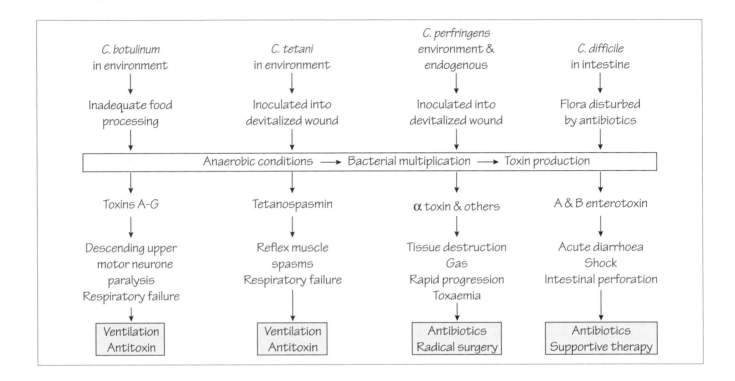

Clostridium spp. are obligate, anaerobic, spore-forming, Gram-positive bacilli. More than 80 species are described, but only a few are human pathogens. All require anaerobic conditions to grow but do vary in their oxygen tolerance and their biochemical profile, e.g. their ability to ferment sugars or digest protein. Their normal habitat is in the soil, in aquatic sediments and in the intestinal tract of both humans and animals. They cause disease as a consequence of toxin production.

Tetanus

Clostridium tetani has both a peritrichous flagella and a large round terminal spore giving the organism a drumstick appearance. A strict anaerobe, it is non-saccharolytic and non-proteolytic. Colonies on blood agar are surrounded by a narrow band of haemolysis.

Epidemiology and pathogenesis

Infection occurs in wounds deep enough to produce anaerobic conditions. The organism produces tetanospasmin which prevents release of the inhibitory transmitter γ-aminobutyric acid (GABA) resulting in muscle spasms. Neonatal tetanus, which may occur if the umbilical stump is contaminated after delivery, is the cause of approximately 500 000 deaths each year.

Generalized tetanus is rare in developed countries (0.2 cases per million), usually occurring in older patients in whom immunity has declined. There is often a history of a trivial gardening injury,

Clinical features

Spastic paralysis and muscle spasms may develop at the site of the lesion; if time allows further toxin production, the condition may become generalized. Then perioral muscle spasm leads to the risus sardonicus ('the sardonic smile'), and spasm of the spinal muscles and legs to opisthotonus (when the head and heels are bent back towards each other). Spasms are painful and may be stimulated by light or sudden noise. There may be respiratory difficulties followed by secondary bacterial pneumonia. Diagnosis is based on history and clinical features: isolation of the organism is not diagnostic.

Treatment and prevention

Treatment aims to reduce symptoms by the use of muscle relaxants and to limit further toxin activity by the use of human tetanus immunoglobulin and toxin production by antibiotics, such as penicillin or metronidazole. Artificial ventilation should be instituted if required, and secondary pneumonia treated appropriately.

Infants are protected by passive immunity from their mothers, and develop active immunity when they receive tetanus toxoid as part of their childhood immunization course. Boosters are given at school entry and every 10–15 years. Unvaccinated patients with tetanus-prone wounds should receive antibiotics and human tetanus immunoglobulin followed by a course of vaccination.

Botulism

Clostridium botulinum has seven types, named A–G, based on biochemical testing and toxin serotype. Serotypes A, B and E are most commonly implicated in human disease.

Epidemiology and pathogenesis

C. botulinum can contaminate foods such as meat or vegetables. Incomplete heat treatment in the canning or bottling process, allows this organism to survive and produce toxin. Botulinum toxin, a neurotoxin, inhibits the release of neurotransmitters. Clinically, there are three forms of the disease: food intoxication, wound botulism and infant botulism. Wound and infant botulism may lead to systemic toxaemia.

Clinical features

A descending flaccid paralysis, beginning with the cranial nerves, develops within 6 h of ingestion of toxin-contaminated food. Patients develop dysphagia and blurred vision followed by more general paralysis but are not confused and sensory function is normal. Infants appear floppy, listless, constipated and have generalized muscle weakness. Diagnosis is based on clinical features and history of suspect food ingestion. Toxin may be demonstrated in faeces and serum by enzyme immunoassay (EIA).

Treatment and prevention

Treatment is with specific antitoxin and ventilatory support. Penicillin is also used to eradicate the organism. Botulinum toxin is inactivated by heating to 80°C for 30 min, and the spores of *C. botulinum* by heating to 121°C for a few minutes. The disease is prevented by adequate process control in the food processing industry and home preservation.

Gas gangrene

Clostridium perfringens is the organism most commonly associated with gas gangrene but *C. septicum*, *C. novyi*, *C. histolyticum* and *C. sordellii* can also be implicated. *C. perfringens* is capsulate and produces a range of toxins of which lecithinase C (α-toxin) is the most important.

Epidemiology and pathogenesis

Typically, gas gangrene develops when a devitalized wound becomes contaminated with spores from the environment. The spores germinate and organisms multiply in the ischaemic conditions, releasing toxins which cause further tissue damage. Progression is rapid.

Clinical features and treatment

Gangrene may develop within 3 days of injury and is associated with considerable pain in the wound. The skin becomes tense and white, with an underlying blue discoloration, foul smell and crepitus. The ensuing toxaemia will produce circulatory shock. The diagnosis is made clinically. Microscopy of stained smears may reveal necrotic material, a few inflammatory cells and large Gram-positive bacteria.

Treatment and prevention

Treatment depends on debridement of devitilized tissue and intravenous antibiotics. Hyperbaric oxygen may also be beneficial. The condition may be prevented by good management of potentially infected devitalized wounds.

Clostridium perfringens food poisoning

This condition is typically associated with meat meals which cool slowly and are reheated. Clostridia in the food release toxin in the stomach which then form spores leading to self-limiting nausea, vomiting and diarrhoea. The diagnosis may be confirmed if >10^5 *C. perfringens* per gram of faeces are isolated. An EIA to detect toxin in faeces is also available.

Pseudomembranous colitis

Clostridium difficile fluoresces on blood-containing media, and isolated using medium containing cycloserine, cefoxitin and fructose.

Epidemiology and pathogenesis

Clostridium difficile is found in the human intestine, especially in those hospitalized patients whose gut microflora has been disturbed by antibiotics. It produces enterotoxins A and B, causing fluid secretion and tissue damage. Neonates commonly carry the organism and toxin without ill-effect; susceptibility increases with age.

Clinical features

Typically, the patient passes more than three loose or unformed stools per day with a history of previous antibiotic exposure. Abdominal pain may develop and sigmoidoscopy will reveal pseudomembranes: small white–yellow plaques situated on the mucosal surface of the rectum and sigmoid colon. The diagnosis is confirmed by the laboratory demonstration of toxin in the stool by EIA or tissue culture.

Treatment and prevention

The inciting agent should be stopped and the patient given oral vancomycin or metronidazole. Patients with pseudomembranous colitis should be isolated from other patients, using enteric precautions.

15 Non-sporing anaerobic infections

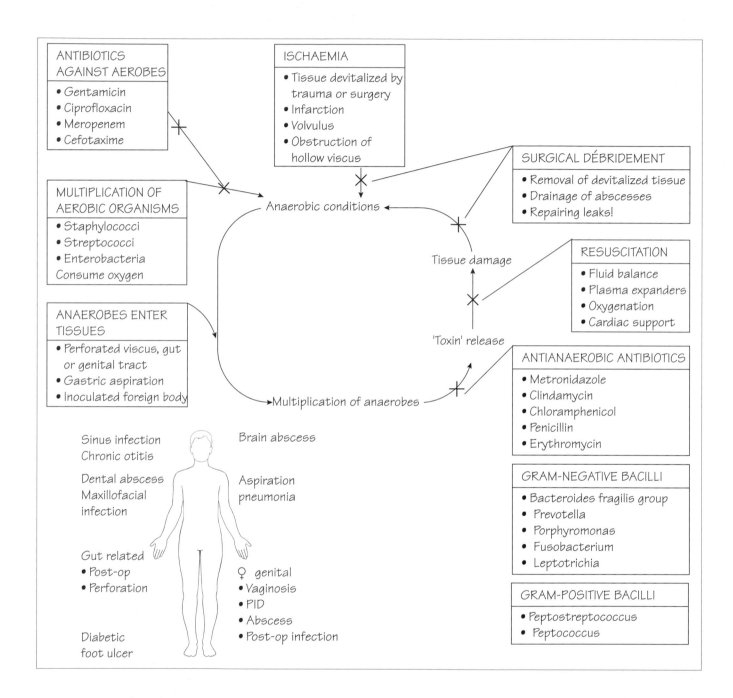

ANTIBIOTICS AGAINST AEROBES
- Gentamicin
- Ciprofloxacin
- Meropenem
- Cefotaxime

ISCHAEMIA
- Tissue devitalized by trauma or surgery
- Infarction
- Volvulus
- Obstruction of hollow viscus

SURGICAL DÉBRIDEMENT
- Removal of devitalized tissue
- Drainage of abscesses
- Repairing leaks!

MULTIPLICATION OF AEROBIC ORGANISMS
- Staphylococci
- Streptococci
- Enterobacteria
Consume oxygen

Anaerobic conditions

Tissue damage

RESUSCITATION
- Fluid balance
- Plasma expanders
- Oxygenation
- Cardiac support

ANAEROBES ENTER TISSUES
- Perforated viscus, gut or genital tract
- Gastric aspiration
- Inoculated foreign body

'Toxin' release

Multiplication of anaerobes

ANTIANAEROBIC ANTIBIOTICS
- Metronidazole
- Clindamycin
- Chloramphenicol
- Penicillin
- Erythromycin

Sinus infection
Chronic otitis

Brain abscess

Dental abscess
Maxillofacial infection

Aspiration pneumonia

Gut related
- Post-op
- Perforation

♀ genital
- Vaginosis
- PID
- Abscess
- Post-op infection

Diabetic foot ulcer

GRAM-NEGATIVE BACILLI
- Bacteroides fragilis group
- Prevotella
- Porphyromonas
- Fusobacterium
- Leptotrichia

GRAM-POSITIVE BACILLI
- Peptostreptococcus
- Peptococcus

Non-sporing anaerobes

Non-sporing anaerobes form the major part of the normal human bacterial flora outnumbering all other organisms in the gut by a factor of 10^3. They are also found in the genital tract, oropharynx and skin.

Anaerobic sepsis

Pathogenesis

Infection with non-sporing anaerobes is endogenous.

Organisms from the normal flora escape into a sterile site, e.g. perforation of the large intestine. Conditions that allow anaerobic growth may arise after ischaemia, e.g. a strangulated hernia, or when facultative bacteria produce anaerobic conditions by the action of their metabolism, e.g. in deep skin ulcers, intraperitoneal infection.

Once established, anaerobic multiplication is promoted by release of toxic metabolic products and proteolytic enzymes. The toxic products of inflammatory cells may also be released, e.g. reactive oxygen intermediates. These exacerbate tissue damage, allowing further anaerobic invasion and multiplication. The cycle of anaerobic sepsis, if not halted, rapidly leads to septicaemia and death.

Clinical importance
Generalized intra-abdominal sepsis may follow spontaneous bowel perforation or elective surgery, leading to abscess formation, e.g. liver abscesses.

Non-sporing anaerobes play a prominent part in sepsis of the female genital tract. These infections are often secondary to septic abortion, prolonged rupture of the membranes, complicated caesarean section or retained products of conception. They are directly implicated in pelvic inflammatory disease, while imbalance in the anaerobic flora of the vagina may lead to the non-specific vaginosis syndrome (see Chapter 44).

Pneumonia following aspiration or associated with carcinoma or foreign body obstruction has a significant anaerobic component. Such infection may develop into a lung abscess.

Brain abscesses often have an important anaerobic component, as does chronic paranasal suppuration, e.g. chronic otitis media and chronic sinusitis.

Anaerobes may colonize chronic skin ulcers, such as lower leg ulceration of the elderly. The less common tropical ulcer is caused by *Fusobacterium ulcerans*.

Laboratory diagnosis
These organisms are nutritionally fastidious: oxygen is toxic to them. Specimens should be plated directly in theatre or at the bedside, or transported to the laboratory rapidly in an anaerobic transport system. Pus rather than swabs (which dry out quickly) should be sent. Specimens are inoculated into a fluid enrichment medium (e.g. Robertson's cooked meat) and onto blood-containing media some of which contain antibiotics to inhibit growth of aerobes. Plates must be incubated anaerobically.

Anaerobic species are identified on the basis of their Gram reaction, their growth on bile- or dye-containing medium, their biochemical reactions, or by studying the end-products of metabolism using gas–liquid chromatography.

Antibiotic susceptibility
Almost all anaerobes are susceptible to metronidazole, although resistance has been reported. Other active agents include meropenem, piperacillin, tazobactam, clindamycin, chloramphenicol, penicillin and erythromycin. (NB Penicillin and erythromycin are not active against *Bacteroides fragilis*, the most common anaerobe isolated.)

Management
Effective management of anaerobic sepsis depends on a dual approach—surgery and antimicrobial agents. Surgical procedures may include closing perforations, resecting gangrenous hernia, débriding non-viable tissue from ulcers, draining abscesses and treating coexisting infection. Metronidazole is the most commonly used antianaerobic agent.

Prevention and control
The risk of anaerobic infection in elective surgery can be reduced by good operative technique and perioperative antibiotics with anti-anaerobic activity.

Pathogens of anaerobic sepsis
• *Bacteroides fragilis* is the most common agent of serious anaerobic sepsis. It is penicillin-resistant by virtue of its β-lactamase production. It also produces a protease, DNAase, heparinase and neuraminidase. It has an antiphagocytic capsule and inhibits the phagocytosis of facultative organisms promoting the development of synergistic infections. *B. fragilis* is typically associated with postoperative sepsis in abdominal and gynaecological surgery. It also contributes to the polymicrobial flora found in cerebral, hepatic and lung abscesses.
• *Prevotella melaninogenicus* and fusobacteria are found chiefly in the oral cavity. They are associated with periodontal disease, gingivitis, dental abscess, sinus infection and cerebral and lung abscesses. They are also found in association with *Borrelia vincenti* in Vincent's angina and in ulcerative diseases, such as cancrum oris (Ludwig's angina), both of which affect the head and neck. They may contribute to anaerobic cellulitis.
• *Peptococcus* and *Peptostreptococcus* are the only anaerobic Gram-positive cocci that are regularly found in human specimens. They are usually found in mixed infections, e.g. dental sepsis, cerebral or lung abscesses, soft tissue and wound infections. They are also associated with necrotizing fasciitis, where a mixed infection of anaerobic cocci, facultative streptococci, and possibly also *S. aureus*, progresses rapidly, with destruction of the skin and deeper tissues, leading to septicaemia and death.

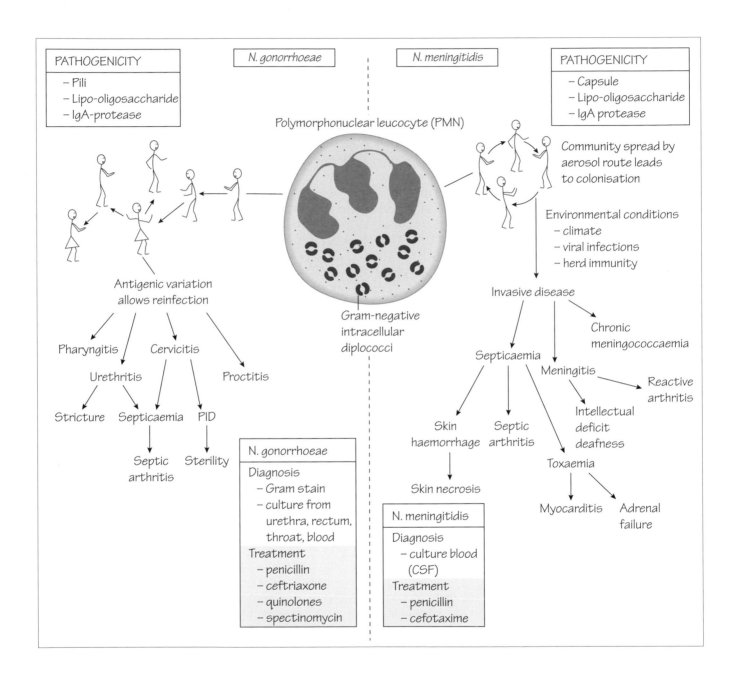

Neisseria gonorrhoeae

Infection with *N. gonorrhoeae* is most common between 15 and 35 years of age. It is almost exclusively spread by sexual contact. Antigenic variation of the gonococcal pili means that recovery from infection provides no immunity.

Pathogenesis

Neisseria gonorrhoeae is a Gram-negative cocci that adheres to the genitourinary epithelium via pili. It invades the epithelial layer and provokes a local acute inflammatory response.

Clinical features

N. gonorrhoeae causes acute painful urethritis and urethral discharge. Female infection (cervicitis) is often asymptomatic or may be associated with vaginal discharge. Pelvic inflammatory disease may develop. Pharyngeal infection causes pharyngitis; rectal infection or proctitis presents with a purulent discharge. Infection can be complicated by bacteraemia, septic or reactive arthritis of the large joints, or pustular skin rashes. Late complications include female infertility and male urethral stricture.

Diagnosis
Pus from urethra, cervix, rectum or throat should plated directly or transported rapidly to the laboratory in specialized transport medium. Identification is based on biochemical tests and serological methods. Expression of β-lactamase can be detected by a rapid colorimetric test.

Treatment and prevention
Treatment must be given before susceptibility results are available. This is usually penicillin: alternatives are ciprofloxacin, ceftriaxone or spectinomycin. Gonorrhoea can be prevented by avoiding high-risk sexual contacts and using barrier contraception. Contacts of infected individuals should be traced and treated. Antigenic variation of the pili precludes vaccine development.

Neisseria meningitidis
Carriage of *N. meningitidis* is common: actual disease only develops in a few individuals. Infection is most common in the winter. Epidemics occur every 10–12 years. In Africa, severe epidemics occur in the 'meningitis belt': here, the incidence can rise to 1000 cases per 100 000 each year. Most invasive infections are caused by serogroup A, B or C. The main pathogenicity determinant of *N. meningitidis* is the antiphagocytic polysaccharide capsule.

Clinical features
Meningococcal meningitis is characterized by fever, neck-stiffness and lowered consciousness. The petechial rash, a sign of septicaemia, may be present without other signs of meningitis. Septic or reactive arthritis may develop.

Treatment
Infection is life-threatening and rapidly progressive: treatment should not await laboratory confirmation or hospitalization. Intravenous penicillin G (intramuscular in the community setting) is the antibiotic of choice, although there have been reports of meningococci with reduced susceptibility in other countries. Cefotaxime is an alternative. Treatment does not eradicate carriage so the patient should be given 'prophylaxis' following recovery.

Prevention
A polysaccharide vaccine is available against serogroups A and C. Protein conjugate vaccines against group C are entering use. Close contacts of patients with meningococcal meningitis should be given 'prophylaxis' with rifampicin or ciprofloxacin.

Moraxella catarrhalis
This Gram-negative cocco-bacillus is usually a commensal of the upper respiratory tract. It is associated with otitis media and, rarely, lower respiratory infections. It produces β-lactamase.

Haemophilus
Haemophilus spp. are small Gram-negative cocco-bacilli, that are dependent for growth on blood factors X and V. They colonize mucosal surfaces. *Haemophilus influenzae* and *H. ducreyi* are the main pathogenic species.

Haemophilus influenzae
Haemophilus influenzae expresses an antiphagocytic polysaccharide capsule of which there are six types (a–f). Septicaemia, meningitis and osteomyelitis are associated with type b. It also expresses a lipopolysaccharide and an IgA_1 protease.

Clinical features
Infection occurs in preschool children causing pyogenic meningitis, acute epiglottitis, septicaemia, facial cellulitis or osteomyelitis. Non-capsulate strains are usually commensal in the nasopharynx but may cause adult otitis media, sinusitis, and chest infection in patients with obstructive airways disease.

Laboratory diagnosis
Antigen detection provides rapid diagnosis in meningitis. Culture of CSF, sputum, blood or pus must be on 'chocolate' agar incubated in 5% CO_2. *H. influenzae* is identified by X and V dependence.

Treatment and prevention
Many *H. influenzae* express a β-lactamase and are ampicillin resistant. Some are also resistant to chloramphenicol. Co-amoxiclav, clarithromycin, tetracycline or trimethoprim can be used. Severe infections are treated with β-lactam stable cephalosporin.

A protein conjugated polysaccharide vaccine has almost eradicated childhood infection. Non-capsulate *Haemophilus* is ubiquitous: predisposed patients cannot avoid infection.

Haemophilus ducreyi
Haemophilus ducreyi is transmitted sexually causing painful, irregular, soft genital ulcers (chancroid). There is associated lymphadenopathy and suppurating inguinal lymph nodes may lead to sinus formation. Infection is more common in developing countries and facilitates the transmission of HIV.

Transmission is controlled by treatment with erythromycin or co-amoxiclav, and contact tracing.

17 Small Gram-negative cocco-bacilli: *Bordetella*, *Brucella*, *Francisella* and *Yersinia*

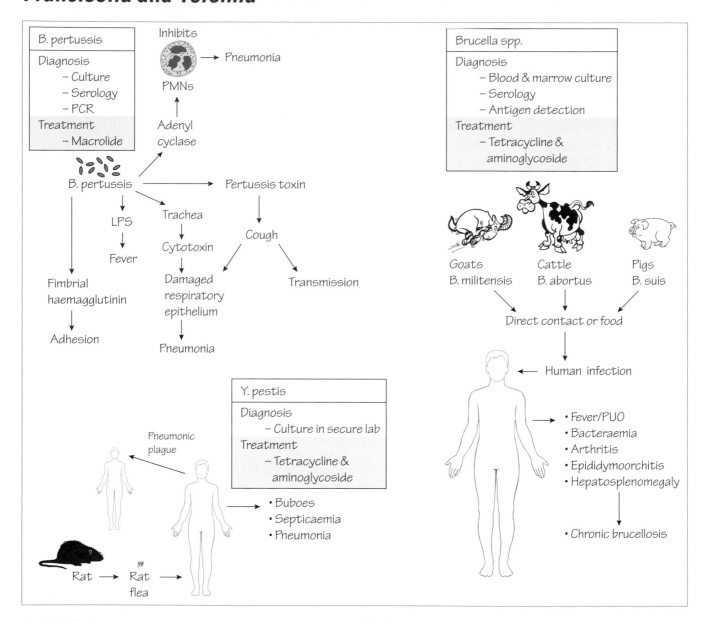

Bordetella spp.

These obligate, aerobic, cocco-bacilli are biochemically inactive. They have fastidious requirements for growth: a medium containing blood, mixed with charcoal to absorb its waste metabolites. *Bordetella pertussis* and *B. parapertussis* cause whooping cough.

In the absence of an adequate vaccination campaign, epidemics of whooping cough occur in children every 4 years. Asymptomatic, or unrecognized infection in adolescents and young adults, maintain the cycle of infection in the human population.

Pathogenesis

Bordetella pertussis expresses fimbriae that aid adhesion, and a number of exotoxins including pertussis toxin, adenyl cyclase and tracheal cytotoxin. There is a complex interaction with the cells of the respiratory tract which results in thickened bronchial secretions and paroxysmal cough. Secondary respiratory tract infection, apnoea induced by prolonged paroxysms of coughing, and raised intracranial pressure are the main complications.

Clinical features

A 2-week catarrhal phase precedes the characteristic whooping cough: repeated, prolonged bouts of coughing followed by a deep inspiratory whoop. The whoop is frequently absent in very young children and adults: it is frequently associated with vomiting and subconjunctival haemorrhage. The coughing phase can last for up to 3 months. Young children feed poorly and lose weight. As spasms are more frequent at night, loss of sleep for parents and child have a significant contribution to the morbidity of the disease. Infection is often complicated by secondary pneumonia and otitis media.

Laboratory diagnosis
Specimens are obtained using a pernasal swab. Inoculated onto charcoal blood agar and incubated for up to 5 days, they develop into minute pearly colonies. Positive cultures are more likely during the catarrhal phase and early coughing phase. Antigen detection using EIA and PCR methods of diagnosis have been described.

Treatment
Erythromycin is thought to decrease infectivity and shorten symptoms if given early during the catarrhal phase. Symptomatic support and early treatment of secondary infections is the mainstay of treatment.

Prevention and control
A whole-cell, killed vaccine is effective when high vaccine coverage of the community is obtained. Long-standing concerns about the safety of this vaccine remain unproved. Subcellular vaccines are used in some countries.

Brucella spp.
Brucella melitensis, *B. abortus* and *B. suis* have goats, cattle and pigs respectively, as their main hosts. They are aerobic or capnophilic, requiring serum-containing medium to grow.

Brucella infection spreads to humans through direct contact with domesticated animals or their products, e.g. unpasteurized milk. Veterinarians, farmers and abattoir workers are at increased risk of infection.

Pathogenesis
Brucellae are able to survive inside cells of the reticuloendothelial system using a superoxide dismutase and nucleotide-like substances to inhibit the intracellular killing mechanisms of the host.

Clinical features
Intermittent high fever is characteristic of the early stages of infection giving rise to the old name, 'undulant fever'. It is associated with myalgia, arthralgia and lumbosacral tenderness. Acute infection can be complicated with septic arthritis, osteomyelitis and epididymo-orchitis. Without treatment a chronic infection develops which may resolve or continue to give symptoms, which are often accompanied by psychiatric complaints, for many years.

Laboratory diagnosis
Culture of blood and bone marrow is diagnostic. Incubation, in a high containment facility, must be extended for up to 3 weeks. In chronic disease, culture is less likely to be positive. Bacterial agglutination is used as a simple screening test: positive results may be confirmed by EIA to detect both IgG and IgM.

Treatment and prevention
Optimal treatment is with tetracycline for 1 month. Streptomycin should be added for patients with complications. Transmission by food can be prevented by pasteurization. Appropriate animal husbandry techniques can reduce the occupational risk of infection. An animal vaccine is available but is not sufficiently safe for human use. Animal control measures have eradicated brucellosis from farms in many countries.

Francisella tularensis
This pathogen of rodents and deer can be found in North America and northern Europe. Infection is spread by the aerosol route, direct contact with feral animals, or by tick bite. This rare infection occurs mainly among campers and hunters. Infection may be ocular or localized to the skin, with regional lymphadenopathy. Systemic infection gives a syndrome that resembles typhoid with 5–10% mortality. Diagnosis is by serology or by culture. Treatment is with tetracycline.

Yersinia
Yersinia pestis
This infection is described in Chapter 48.

Yersinia enterocolitica
This organism, morphologically and biochemically similar to *Y. pestis*, causes acute enteritis, mesenteric adenitis and, rarely, septicaemia. It is transmitted to humans in food and water. Infection may be complicated by polyarthritis and erythema nodosum. Patients with iron overload syndromes are especially susceptible. Diagnosis is made by isolation from faeces, blood or lymph node, or detection of antibodies. Treatment with ciprofloxacin or co-trimoxazole is indicated in serious infection; tetracycline is an alternative therapy.

Yersinia pseudotuberculosis
This organism can cause mesenteric adenitis that mimics appendicitis.

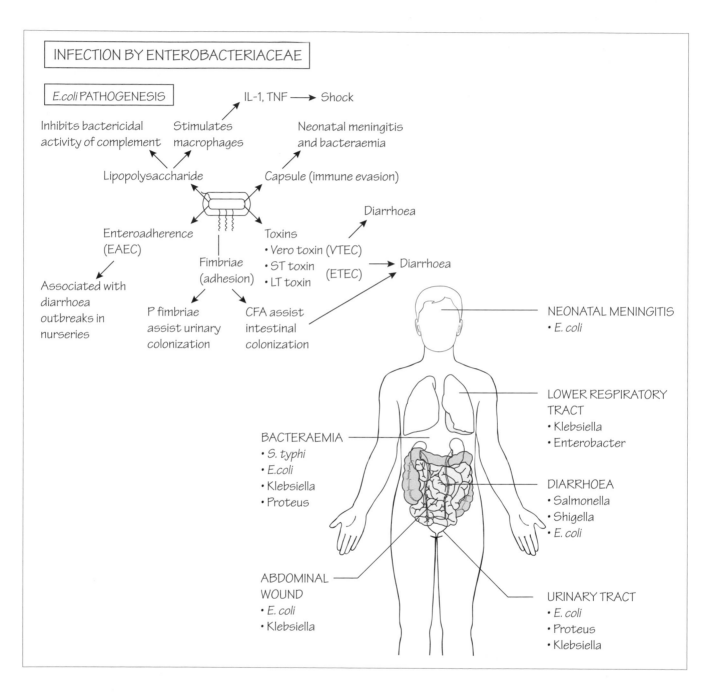

The Enterobacteriaceae are a large family (more than 20 genera and 100 species) of facultatively anaerobic, Gram-negative bacilli that are easily cultured, reduce nitrate, and ferment glucose.

Pathogenicity
Capsules
Many produce extracellular capsular polysaccharides, e.g. *Klebsiella* spp., *Escherichia coli* and *Salmonella typhi*.

S. typhi possesses a capsule or Vi (virulence) antigen: vaccine containing the Vi antigen is protective. *Escherichia coli* K1 is the most common type of *E. coli* isolated from neonatal meningitis and septicaemia.

Lipopolysaccharide
The lipopolysaccharide (LPS) molecule consists of a central lipid A and oligosaccharide core, and a long straight or branched polysaccharide 'O' antigen. It is located in the

bacterial outer membrane and is responsible for resistance to the bactericidal activity of complement. The lipid A core stimulates host macrophages to produce cytokines, e.g. interleukin-1 and tumour necrosis factor (TNF), responsible for the fever, shock and metabolic acidosis associated with severe sepsis. Some clinical syndromes are associated with different O antigens, e.g. *E. coli* O157 may produce verotoxin causing haemolytic–uraemic syndrome.

Urease
Proteus spp. express a potent urease that splits urea, thus lowering urinary pH. This is associated with formation of urinary stones (see Chapter 44).

Fimbriae
Fimbriae are bacterial organelles that allow attachment to host cells and are important in promoting colonization. *Escherichia coli*, expressing mannose-binding fimbriae, are associated with lower urinary tract infections and cystitis, whereas those that express P fimbriae are associated with pyelonephritis and septicaemia. In the intestine, *E. coli* that express different fimbriae (colonization factor antigens, CFA) have been associated with diarrhoea.

Toxins
Verotoxin
This toxin is so named because of its activity on vero cells. Toxin-producing strains of *E. coli* are known as enterohaemorrhagic *E. coli* (EHEC). The haemorrhagic diarrhoeal syndrome that results can be complicated by haemolysis and acute renal failure: the haemolytic–uraemic syndrome. This organism is commensal in cattle and is transmitted to humans through hygiene failure in abattoirs and food production.

Shiga toxin
Some strains of *Shigella* and *E. coli* are able to produce a toxin known as the Shiga toxin.

LT toxin and ST toxin
These toxins act on the enterocyte to stimulate fluid secretion resulting in diarrhoea. LT toxin is heat labile and increases local cyclic adenosine monophosphate (cAMP) in the enteric cell. ST is heat stable and stimulates cyclic guanylyl monophosphate. *Escherichia coli* possessing these enterotoxins is associated with travellers' diarrhoea: a short-lived, watery diarrhoeal disease.

Enteroaggregative *Escherichia coli*
Some strains of *E. coli* are able to attach to, and cause aggregation of, enteric cells. These are known as enteroaggregative *E. coli* (EAEC) and are able to cause chronic diarrhoea. The molecular basis of this pathological process is not yet understood.

Enteroadherent *Escherichia coli*
The *E. coli* with this characteristic were the first *E. coli* recognized as primary pathogens causing outbreaks of diarrhoea in pre-school nurseries.

Clinical syndromes
These organisms cause urinary tract infections (see Chapter 44), diarrhoea (see Chapter 46) and bacteraemia (see Chapter 40).

Enteric fever
Enteric fever (typhoid) is caused by *Salmonella typhi* or *S. paratyphi*. Invasion of the intestinal wall, with spread to local lymph nodes, is followed by primary bacteraemia and infection of the reticuloendothelial system. The bacteria reinvade the bloodstream and gut from the gallbladder, multiply in Peyer's patches, and cause ulceration that may be complicated by haemorrhage or perforation. Patients present with fever, alteration of bowel habit (diarrhoea or constipation) and the classical but rare rash (rose spots on the abdomen). Hepatosplenomegaly may also be demonstrated. Enteric fever may be complicated by osteomyelitis and, rarely, by meningitis.

Diagnosis
Enteric Gram-negative organisms are identified by biochemical reactions, e.g. the pattern of fermentation of different sugars. Epidemiological investigation uses serotyping (sera directed against the LPS (O) antigens and flagellar (H) antigens), phage typing or colicine typing (using the pattern of inhibition produced by these proteins).

A diagnosis of typhoid is made by isolating organisms from the blood or bone marrow.

Treatment and prevention
Most enteric Gram-negative organisms are susceptible to aminoglycosides, third-generation cephalosporins, fluoroquinolones, β-lactams and carbapenems (e.g. meropenem). As some produce β-lactamases and aminoglycoside-degrading enzymes, treatment should be guided by sensitivity tests.

In urinary tract infections, cefalexin, ampicillin or trimethoprim are the antibiotics of first choice.

Diarrhoeal disease can be avoided by good hygiene, food preparation and safe water supplies. Treatment is primarily by oral rehydration (see Chapter 46).

Ciprofloxacin is the treatment of choice for typhoid; alternatives are trimethoprim or third-generation cephalosporins. Multidrug-resistant typhoid has been a major problem in some countries. A live attenuated vaccine (Ty21A) or a subcellular vaccine (containing the Vi antigen) is available for travellers to areas of high risk, but gives only partial protection.

19 *Vibrio, Campylobacter* and *Helicobacter*

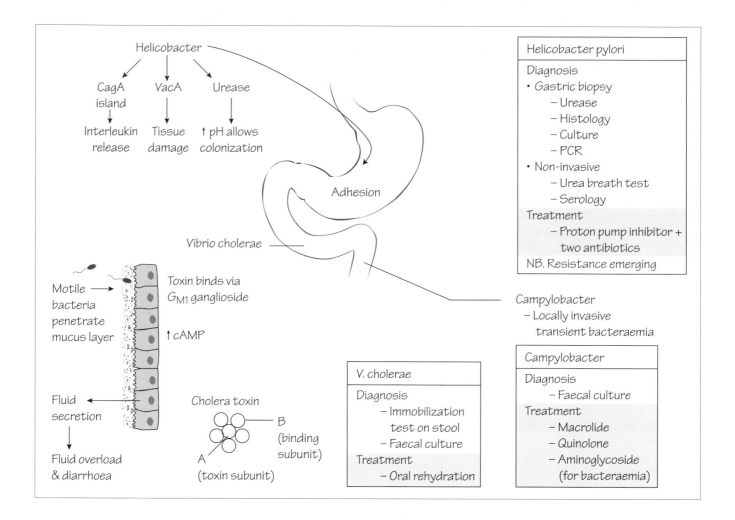

Vibrio spp.

Vibrios are small, Gram-negative, curved bacilli, that are motile and oxidase-positive. There are more than eight species. *Vibrio cholerae* and *V. parahaemolyticus* are the main human pathogens.

Vibrio cholerae

The organism is subdivided by the somatic O antigens: only O1 and O139 are associated with disease. It survives gastric acid, burrows through the intestinal mucus to attach to intestinal epithelial cells via the G_{M1} ganglioside and produces a multimeric protein toxin (cholera toxin). This stimulates adenyl cyclase within the cell resulting in the secretion of water and electrolytes into the lumen of the bowel.

Epidemiology

Cholera is an exclusively human illness. The disease is trans-
mitted via water and food, especially seafood. It is primarily a disease of developing countries where there is inadequate sanitation and safe water supplies. Epidemic disease appears to follow a change in the environment, but is facilitated by human behaviour, e.g. war, refugee movements and migration. Cholera gives rise to periodic pandemics: the present epidemic is the seventh recorded.

Clinical features

Many infections with *V. cholerae* are mild or asymptomatic. The cholera syndrome is characterized by massive, painless, fluid diarrhoea up to 20 litres per day, which may be accompanied by vomiting. Severe dehydration follows, complicated by electrolyte imbalance.

Diagnosis

Where cholera is endemic, the diagnosis is based on clinical features. Immobilization of cholera bacteria in a diarrhoeal stool with specific antiserum, yields a rapid diagnosis. Antigen and molecular tests have been described but are inappropriate in the countries where cholera is common. The organism can be cultivated on selective medium (thiosulphate citrate bile-salt sucrose medium, or enriched in alkaline peptone water). Biochemical identification and serotyping should be performed to confirm the diagnosis.

Treatment

Oral rehydration solution (salt and glucose mix) is very effective, although intravenous fluids may be required for severely ill patients. Antibiotics, e.g. tetracycline or cipro-floxacin, can shorten the duration and severity.

Prevention and control

Provision of safe water supplies and community education does much to prevent epidemic cholera. Several experimental live attenuated and subunit vaccines are under trial.

Campylobacter spp.

Campylobacter are microaerophilic, curved, Gram-negative rods; they are motile by virtue of polar flagella. They are associated with diarrhoeal disease and cause infection more commonly than *Salmonella* and *Shigella*. Although there are more than 18 species of *Campylobacter*, *C. jejuni* alone is responsible for 90% of *Campylobacter* gastrointestinal infections. Infection follows ingestion of contaminated meat, poultry, unpasteurized milk or contaminated water. *Campylobacter coli* causes bacteraemia in compromised patients.

Pathogenicity

C. jejuni invades and colonizes the mucosa of the small intestine.

Clinical features

Patients typically complain of influenza-like symptoms, crampy abdominal pain and diarrhoea which may be blood-stained. Among children, the pain may be sufficiently severe to suggest appendicitis or intussusception. A self-limiting bacteraemia is common. Guillain–Barré syndrome is associated with *Campylobacter* infection.

Diagnosis

Faecal samples should be inoculated onto a *Campylobacter*-specific medium containing lysed blood and a mixture of antibiotics, and incubated at 42°C in a microaerophilic atmosphere. Identification of bacteria is based on their growth at 42°C, their characteristic microscopic morphology ('seagull wing') and expression of catalase and oxidase.

Treatment

Diarrhoea is often self-limiting but patients may be treated with erythromycin or fluoroquinolones. An aminoglycoside may be added for patients who have septicaemia.

Prevention and control

Prevention of campylobacteriosis depends on good animal husbandry and abattoir practices, and good food hygiene in shops, dairies and the home.

Helicobacter pylori

Helicobacter pylori is a Gram-negative, non-sporing, microaerophilic, spiral bacillus which is motile by virtue of five or six unipolar flagella. It is catalase-, oxidase- and urease-positive.

Helicobacter cinaedi and *H. fennelliae* have been isolated from HIV-positive individuals with proctocolitis and with bacteraemia.

Pathogenesis

Helicobacter pylori expresses urease, raising the pH in the surrounding locality, and thus protecting the bacterium from the effects of gastric acid. The CagA protein (120–140 kDa) is found in organisms isolated from peptic ulcer patients. Its function is unknown but it is immunogenic, stimulating interleukin-8 production. In addition, the organism also expresses lipases, haemolysins and the vacuolating cytotoxin VacA, a secreted protein which damages cells.

Clinical features

Infection, whether acute or chronic, is often asymptomatic. Chronic infection often takes the form of a low-grade gastritis. There is a very strong association with both gastric and duodenal ulceration, and less so with gastric cancer.

Diagnosis

Infection may be confirmed by gastric and duodenal biopsy at endoscopy. The biopsy is examined histologically, by microbiological culture and PCR. Demonstration of urease activity in the biopsy suggests the diagnosis. An infected patient given an oral dose of ^{13}C- or ^{14}C-labelled urea will excrete labelled CO_2 in the breath, which can be detected. The diagnosis can also be made by detecting antibodies in serum by EIA.

Treatment

A combination of antibiotics and a proton pump inhibitor (triple therapy) appears to be most successful, e.g. amoxicillin, metronidazole and omeprazole. Other therapeutic variations are equally valid, but the ideal combination is not yet clear. Reinfection with *H. pylori* can occur.

20 Environmental pathogens: *Pseudomonas, Burkholderia* and *Legionella*

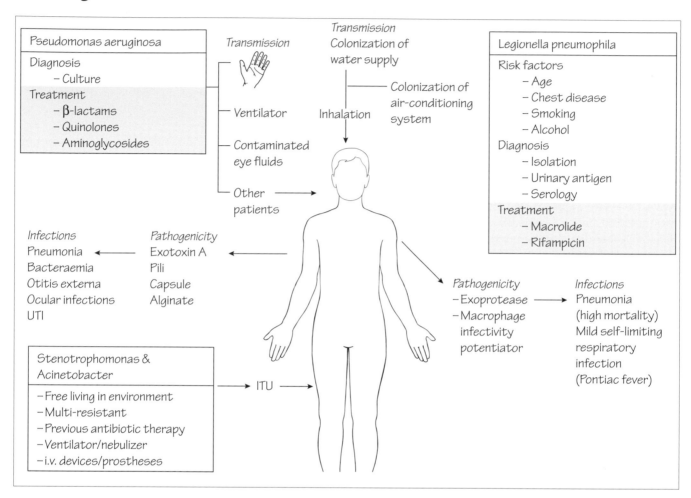

Pseudomonas spp.

The genus *Pseudomonas* are environmental organisms which may cause opportunistic infections. Most cases are attributed to *Pseudomonas aeruginosa*.

Pseudomonas aeruginosa

This organism is a motile, aerobic, Gram-negative bacillus. It is ubiquitous in the environment but rare in the flora of healthy individuals. Carriage increases with hospitalization. Moist environments harbour *P. aeruginosa*, e.g. sink traps, taps and disinfectants in use over 24 h.

Pathogenesis

Pseudomonas aeruginosa produces cytotoxins and proteases, e.g. exotoxin A and S, haemolysins and elastase. Isolates from patients with cystic fibrosis produce a polysaccharide alginate. This allows microcolonies to form where organisms are protected from opsonization, phagocytosis and antibiotics. The alginate, pili and outer membrane protein mediate adherence. Alginate production is associated with hypersusceptibility to antibiotics, LPS deficiency, non-motility and reduced exotoxin production.

Clinical features

Corneal infection can be rapidly progressive as can *Pseudomonas* otitis externa. Burns can become colonized leading to secondary septicaemia. Septicaemia with a high mortality is a particular threat to neutropenic patients. A few show a destructive skin complication, ecthyma gangrenosum. Osteomyelitis, septic arthritis and meningitis can occur, the latter usually after neurosurgery. Chronic infection of cystic fibrosis patients causes a progressive deterioration of lung function.

Laboratory diagnosis

Pseudomonas aeruginosa grows on most media but those containing cetrimide, irgasin and naladixic acid are selective. The organism is identified by biochemical testing and its ability to grow at 42°C. It may be typed by O and H agglutination reactions, bacteriophage typing, bacteriocin typing or molecular methods, e.g. pulse-field gel electrophoresis.

Treatment
Treatment is with aminoglycosides, carbapenems, ureidopenicillins, expanded spectrum cephalosporins or fluoroquinolones. Organisms may exhibit multiple resistance.

Prevention and control
Vaccination is not effective. Spread of multi-resistant strains should be controlled within hospitals by isolating infected individuals and reducing moist environments in which the organism may colonize.

Burkholderia spp.
Burkholderia cepacia
This organism causes chronic pulmonary infection among cystic fibrosis patients. It leads to a decline in pulmonary function or fulminant septicaemia. It spreads from person to person in cystic fibrosis clinics. Naturally resistant to many antibiotics, treatment is with expanded spectrum cephalosporins, carbapenems or ureidopenicillins and is based on susceptibility testing.

Burkholderia pseudomallei
This tropical organism is a free-living saprophyte of soil and water. It causes melioidosis, which presents as a tuberculosis-like disease, as acute septicaemia or as multiple abscesses. Septicaemia is associated with a high mortality. The diagnosis is made by cultivating the organism from blood or tissues. Treatment is with ceftazidime. *Burkholderia mallei* causes a similar infection in horses, known as glanders, which can spread to humans.

Stenotrophomonas maltophilia
Stenotrophomonas maltophilia is a Gram-negative bacillus normally found in soil and water. It inhabits moist environments and, because it is resistant to many antibiotics, can colonize patients in ITU and the immunocompromised. Infection is transmitted by staff and by contaminated shared equipment, such as nebulizers. The organism causes septicaemia or pneumonia. Most strains are resistant to aminoglycosides and carbapenems but are sensitive to co-trimoxazole, tetracycline and expanded spectrum cephalosporins.

Acinetobacter spp.
Acinetobacter are small Gram-negative cocco-bacilli. They are environmental organisms, naturally resistant to many antibiotics and colonize patients in hospitals, especially those in ITU. They can colonize the inanimate environment in damp places, such as humidifiers, and are implicated in outbreaks of multidrug-resistant infection. Systemic invasion leads to pneumonia, septicaemia, meningitis or urinary tract infection. Infection is more likely in patients receiving antibiotics, with multiple cannulae or who are intubated. Treatment, when indicated, is based on the results of susceptibility tests.

Legionella spp.
Legionellae are fastidious, Gram-negative, pleomorphic bacteria. There are more than 39 species but only *L. pneumophila* is regularly implicated in human disease. *Legionella* spp. are found in rivers, lakes, warm springs and domestic water supplies, fountains, air-conditioning systems, swimming pools and jacuzzis. The organisms multiply in water between 20 and 40°C often in association with other microorganisms, e.g. cyanobacteria, or *Acanthamoeba*. They are transmitted to humans when aerosols are generated and inhaled, e.g. in showers and air-conditioning systems. Legionnaires' disease is associated with previous lung disease, smoking and high alcohol intake, but previously healthy patients can be infected. Immunocompromised patients in hospital are vulnerable to infection if the hospital air-conditioning system is not adequately maintained.

Pathogenesis
Pathogenicity factors include the major outer membrane protein that inhibits acidification of the phagolysosome and the macrophage infectivity potentiator that is required for optimal internalization. *Legionella pneumophila* expresses a potent exoprotease.

Clinical features
Legionellosis may take the form of a mild influenza-like illness. Equally, pneumonia (Legionnaires' disease) can be severe with respiratory failure and high mortality. Patients often complain of gastrointestinal symptoms, e.g. nausea or vomiting and malaise, before lung symptoms become prominent. The cough is usually unproductive but dyspnoea is progressive. Mental confusion is common.

Laboratory diagnosis
Sputum, or preferably bronchoalveolar lavage fluid, should be cultured. Suspect colonies are identified serologically. Rapid diagnosis is by direct immunofluorescence or PCR of respiratory specimens and antigen detection in urine.

Treatment and prevention
Effective regimens usually consist of a macrolide antibiotic together with rifampicin.

Legionellosis is prevented by maintenance of air-conditioning systems and ensuring that the hot water supply is above 45°C to prevent multiplication.

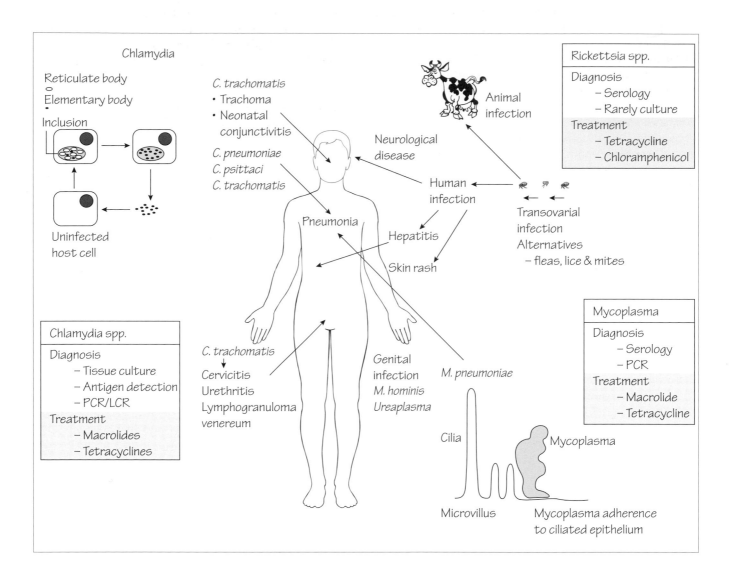

Chlamydia

There are three species: *C. trachomatis*, which infects the eye and the genital tract, and two respiratory pathogens, *C. psittaci* and *C. pneumoniae*. They are obligate, intracellular bacteria which exist in two forms: the reticulate body (non-infective intracellular vegetative form) and the elementary body (the extracellular form permits the organism to survive and be transmitted) which is derived from the reticulate body by binary fission.

Pathogenicity

The major outer membrane protein may participate in attachment to mucosal cells. A 60 kDa cysteine-rich protein may also be associated with virulence.

Chlamydia psittaci

Chlamydia psittaci, a pathogen of birds and mammals, causes psittacosis. It is transmitted from birds to humans. After an incubation period of 10–14 days, fever, a dry unproductive cough, dyspnoea, headache and myalgia develop. Clinical examination reveals few signs of consolidation but chest X-ray may show patchy consolidation.

Chlamydia pneumoniae

Chlamydia pneumoniae is transmitted from person to person by the respiratory route. It produces a pneumonia or bronchitis which is clinically mild but may be associated with pharyngitis, sinusitis and laryngitis. There is increasing evidence that *C. pneumoniae* is associated with the development of atherosclerosis.

Chlamydia trachomatis
There are 13 serotypes: A–C are associated with trachoma (see Chapter 49), neonatal conjunctivitis (see Chapter 38) and D–K associated with acute urethritis and pelvic inflammatory disease (see Chapter 44). Serotypes L1–L3 are associated with lymphogranuloma venereum.

Laboratory diagnosis
Psittacosis is usually diagnosed serologically using species-specific microimmunofluorescence, IgM tests, and IgM capture EIA.

Chlamydia trachomatis is readily cultivated but antigen detection EIA, PCR or ligase chain reaction is more commonly used for diagnosis.

Chlamydia pneumoniae can be grown in HeLa cells. Complement fixation tests (CFTs), EIAs and PCR-based methods are entering routine diagnosis.

Mycoplasma and *Ureaplasma*
Mycoplasma and *Ureaplasma* are small bacteria that lack a cell wall. They are parasites of animals, arthropods and plants. *Mycoplasma pneumoniae* is a primary human pathogen. *Mycoplasma hominis* and *U. urealyticum* are commensal but may be associated with human infection.

Mycoplasma pneumoniae
This is an important cause of atypical pneumonia. After *S. pneumoniae*, it is probably the second most common cause of acute, community-acquired pneumonia.

Pathogenicity
Mycoplasma pneumoniae adheres to host cells by the P1-protein, a 169 kDa antigen. Immunity is short-lived: antigenic variation in the P1-protein is responsible for this. *M. pneumoniae* locates itself at the base of the cilia where it induces ciliostasis. Secreted hydrogen peroxide damages host membranes and interferes with superoxide dismutase and catalase. Opsonized *M. pneumoniae* is readily killed by macrophages and by the activity of the complement system.

Clinical features
Patients present with fever, myalgia, pleuritic chest pain and a non-productive cough; headache is a prominent symptom. Antibodies, that agglutinate the host's own red cells at low temperature, result in peripheral and central cyanosis after exposure to the cold. Infection is associated with reactive (postinfective) arthritis, and neuritis.

Laboratory diagnosis
Culture growth of either *M. pneumoniae* or *U. urealyticum* is too slow to be of clinical value. A four-fold rise in CFT between acute and convalescent specimens indicates acute infection. An IgM-specific EIA, using a P1-protein surface antigen, is more sensitive than CFT, giving a positive result on a single specimen.

Treatment
These organisms are resistant to β-lactams and cephalosporins but are sensitive to erythromycin, tetracycline, aminoglycosides, rifampicin, chloramphenicol and quinolones.

Rickettsia
These organisms are obligate intracellular bacteria with biochemical similarities to Gram-negative bacteria. Clinically, they are divided into three groups:
1 Rocky Mountain spotted fever (RMSF)
2 Spotted fever
3 Typhus

The typhus group includes *R. prowazeki* and *R. typhi* which cause epidemic and murine typhus, respectively; the scrub typhus group is caused by a single species, *R. tsutsugamushi*.

The incubation period is up to 14 days. After non-specific symptoms, patients may develop fevers, arthralgia and malaise followed by the development of a rash, conjunctivitis and pharyngitis. Confusion only occurs in a proportion of RMSF cases. Relapse of *R. prowazeki* infection months or years later is known as Brill–Zinsser disease but is usually milder than the primary infection.

Diagnosis is by immunofluorescence, CFT, IgM-specific EIA or PCR. Tetracyclines and chloramphenicol are the treatment of choice, but must be initiated early to influence the outcome.

Coxiella burneti
Coxiella burneti is a small, Gram-negative, rod-shaped bacteria closely related to *Rickettsia*. It usually infects cattle, sheep and goats, localizing in the placenta. It survives desiccation in the environment and is transmitted predominantly by the aerosol route.

Coxiella burneti is the causative organism of Q fever which may present as an atypical pneumonia or pyrexia of uncertain origin. In about 50% of infected patients the clinical picture is dominated by hepatitis and splenomegaly. Relapses take the form of culture-negative endocarditis or granulomatous hepatitis.

Q fever is usually diagnosed by CFT using acute and convalescent serum. *Coxiella* express different antigens at different phases of infection. EIA and PCR-based methods are available in reference laboratories.

22 Spiral bacteria

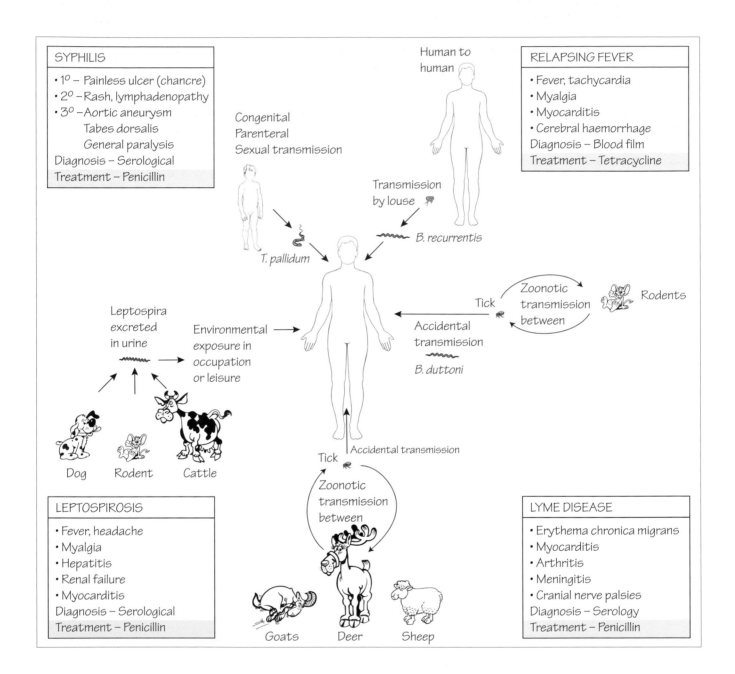

SYPHILIS
- 1° – Painless ulcer (chancre)
- 2° – Rash, lymphadenopathy
- 3° –Aortic aneurysm
   Tabes dorsalis
   General paralysis
- Diagnosis – Serological
- Treatment – Penicillin

Human to human

RELAPSING FEVER
- Fever, tachycardia
- Myalgia
- Myocarditis
- Cerebral haemorrhage
- Diagnosis – Blood film
- Treatment – Tetracycline

Congenital
Parenteral
Sexual transmission

Transmission by louse

B. recurrentis

T. pallidum

Leptospira excreted in urine

Environmental exposure in occupation or leisure

Tick

Zoonotic transmission between

Rodents

Accidental transmission

B. duttoni

Dog   Rodent   Cattle

Tick   Accidental transmission

Zoonotic transmission between

LEPTOSPIROSIS
- Fever, headache
- Myalgia
- Hepatitis
- Renal failure
- Myocarditis
- Diagnosis – Serological
- Treatment – Penicillin

Goats   Deer   Sheep

LYME DISEASE
- Erythema chronica migrans
- Myocarditis
- Arthritis
- Meningitis
- Cranial nerve palsies
- Diagnosis – Serology
- Treatment – Penicillin

Leptospira

Leptospira are aerobic, tightly coiled, motile bacteria. There are only two species: *L. interrogans* and *L. biflexa*—a non-pathogen. *L. interrogans* has more than 200 serovariants, e.g. *L. interrogans* var. *icterohaemorrhagiae* shortened to *L. icterohaemorrhagiae*.

Epidemiology

Leptospira are parasitic, with different preferred mammalian hosts, e.g. the rat is the reservoir of *L. icterohaemor-*

rhagiae. Leptospires colonize the renal tubules of their natural host and are excreted in the urine. Humans may be infected from contact with animal urine or by contaminated water or soil. Water sports enthusiasts and agricultural and abattoir workers are at increased risk of infection.

Pathogenesis and clinical features

The central nervous system, liver and kidneys are most affected in human disease; the severity varies between serovars, e.g. *L. icterohaemorrhagiae* infection is usually more severe than *L. hardjo* infection.

Leptospirosis has two phases: bacteraemia, with fever, headache, myalgia, conjunctivitis and abdominal pain; after the organisms disappear from the blood, fever, uveitis and aseptic meningitis predominate. Jaundice, haemorrhage, renal failure and myocarditis occur in severe cases associated with a significant mortality.

Diagnosis and treatment

Leptospira can be cultured from the blood during the first week of illness. Rising antibody titres can be detected by microagglutination techniques or IgM-specific EIA.

Penicillin or tetracycline must be commenced early in the disease. Tetracycline is an effective prophylactic agent if exposure to infection is likely to have occurred.

Borrelia

Borrelia are loosely coiled, spiral bacteria. They are transmitted to humans via arthropods: lice or ticks. Infections arise sporadically throughout the world with a well-defined geographical territory and host specificity, e.g. humans are the only host of louse-borne relapsing fever (*B. recurrentis*). Epidemics arise during war or mass migration when humans invade the *Borrelia*–tick–rodent habitat.

Relapsing fever

Borrelia invade the bloodstream, producing fever. Antibodies clear the organism from the blood, but antigenic variation allows relapse. Untreated, relapsing fever resolves when the organism exhausts its repertoire of antigenic variation.

Patients experience headaches, myalgia, tachycardia and rigors; examination may reveal hepatosplenomegaly and a petechial rash. Episodes last for 3–6 days; relapses are approximately a week apart. Louse-borne relapsing fever has a high mortality (up to 40%); tick-borne disease mortality rarely exceeds 5%. Dysrhythmias (secondary to myocarditis), cerebral haemorrhage or hepatic failure are the usual causes of death.

Lyme disease

Borrelia burgdorferi causes Lyme disease: it is transmitted by *Ixodes* ticks. Endemic in the eastern United States and Europe, humans are accidental hosts. Localized skin infection is followed by migration through the skin and dissemination throughout the body. The early symptoms are caused by the acute infective process; later manifestations are thought to be related to the host immune response.

Initially an expanding red macule or papule (erythema chronicum migrans) may be found. This is followed by headache, conjunctivitis, fever and regional lymphadenopathy. New skin lesions, myocarditis, arthritis, aseptic meningitis, cranial nerve palsies and radiculitis are complications.

Acrodermatitis chronica atrophicans, a red skin lesion, may also occur.

Vincent's angina

This is a painful, ulcerative, synergistic infection between *Borrelia vincenti* and fusobacteria in the mouth.

Diagnosis and treatment

Diagnosis is by visualizing *Borrelia* in the peripheral blood. Lyme disease can be diagnosed using specific EIA.

Tetracycline or amoxicillin is used for treatment of early Lyme disease: ceftriaxone for late or recurrent disease. Tetracycline is the drug of choice for treating relapsing fever.

Treponema pallidum

The causative organism of syphilis, *Treponema pallidum*, is transmitted sexually and congenitally. The characteristic syphilitic lesions (gumma)—necrosis and obliterative endarteritis with fibroblastic proliferation and lymphocyte infiltration—are found throughout the body.

Clinical features

Organisms penetrate intact skin and then disseminate throughout the body. There are four phases of disease: primary chancre (painless ulcer with rubbery edge and regional lymphadenopathy); secondary (an acute febrile illness with a generalized non-itchy scaling rash, typically involving the palms, and associated with lymphadenopathy); a long latent phase lasting many years; tertiary (systemic lesions become symptomatic, e.g. aortitis, posterior cord degeneration and dementia).

Diagnosis

T. pallidum can be seen by dark-ground microscopy in specimens from the primary chancre or rash. Serology uses EIA for specific IgG and IgM. Alternative tests use cardiolipin agglutination (measures disease activity) and tests based on cultivated treponemes (treponemal haemagglutination test). CSF testing should be performed to detect early central nervous system involvement.

Treatment

Syphilis is treated with penicillin (or tetracyclines if allergic). An acute febrile response (the Jarisch–Herxheimer reaction) may develop in some patients after the first dose of antibiotics. Careful serological follow-up is required to confirm cure and detect early central nervous system involvement.

Related organisms, *T. pertenue* and *T. carateum*, cause yaws and pinta respectively. They are non-venereal and spread by contact, usually in childhood. Once common in the tropics, they are rare as a result of an eradication campaign.

23 Virus structure, classification and antiviral therapy

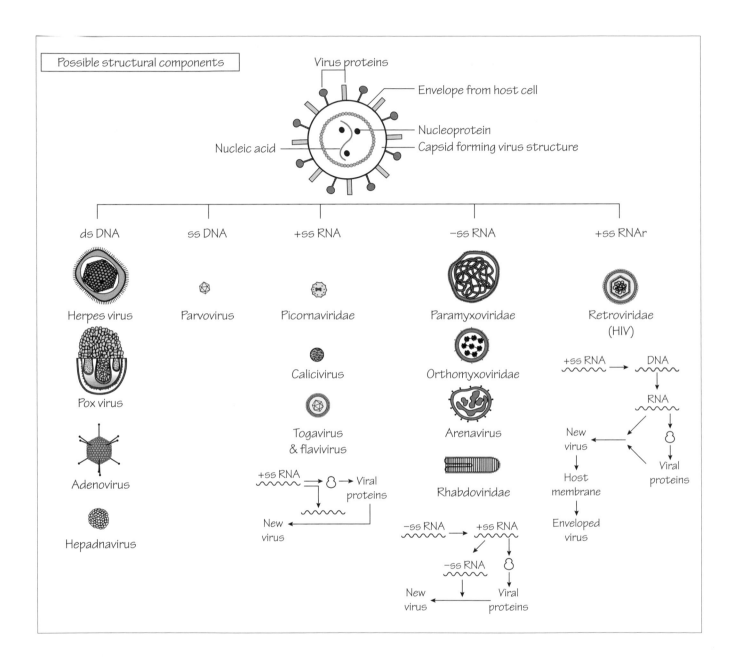

Viral classification

Modern classification of viruses is based on the genetic material, the mode of replication, the structure and symmetry of the structural proteins (capsids) and the presence of an envelope.

Genetic material and replication

DNA viruses

Virus DNA is either double (ds) or single stranded (ss). Double-stranded DNA viruses include a number of important species that cause human disease: poxviruses; herpesviruses; adenoviruses; papovaviruses and polyomaviruses. These last two are small viruses associated with benign tumours, such as warts and malignant tumours, e.g. cervical cancer. Hepatitis B virus is double stranded with single-stranded portions. Among the single-stranded DNA viruses, parvoviruses are responsible for erythema infectiosum.

DNA viruses usually replicate in the nucleus of host cells producing a polymerase which reproduces viral DNA. Viral DNA is not usually incorporated into host chromosomal DNA.

RNA viruses

RNA viruses possess a single strand of RNA and adopt different reproduction strategies depending on whether the RNA is sense or antisense. RNA sense (positive) may serve directly as mRNA. It is translated into structural protein and an RNA-dependent RNA polymerase. A virus with RNA antisense (negative) contains an RNA-dependent RNA polymerase that transcribes the viral genome into mRNA. Alternatively, the transcribed RNA can act as a template for further viral (antisense) RNA.

Retroviruses possess single-stranded sense RNA that cannot act as mRNA. It is transcribed into DNA by reverse transcriptase. This DNA is incorporated into host DNA. The subsequent transcription to make mRNA and viral genomic RNA, is under the control of host transcriptase enzymes.

Capsid symmetry

Viral nucleic acid is covered with a protein coat made up of repeating units (capsids), which exhibit one of two forms of symmetry: icosahedral or helical. In icosahedral symmetry, the capsids form an almost spherical structure. Helical symmetry is found in RNA viruses that have capsids bound around the helical nucleic acid. A structure based on repeating units reduces the number of genes devoted to the viral coat and simplifies the process of viral assembly.

Envelope

In some viruses the nucleic acid and capsid proteins of the virus (the nucleocapsid) are surrounded by a lipid envelope derived from the host cell or nuclear membranes. The host membrane is altered by viral encoded proteins or glycoprotein, which may act as receptors for other host cells. Enveloped viruses are sensitive to substances that dissolve the lipid membrane.

Antiviral therapy

The intracellular location of viruses and their use of host cell systems makes antiviral therapy difficult to develop. Despite this, there are a growing number available: some of the more important are listed below.

Amantadine

Amantadine is active against influenza A. It prevents viral uncoating and release of viral RNA. Resistance is documented. Short courses may prevent disease during outbreaks but are usually reserved for patients at high risk.

Nucleoside analogues

Aciclovir is phosphorylated by virally encoded thymidine kinase. This is not a human enzyme but only occurs in virally infected cells. As it is active against herpesviruses and varicella-zoster virus, it is used to treat herpes simplex infections and for prophylaxis against herpes infections in the immunocompromised. Resistance occurs through the development of deficient thymidine kinase production or alteration in the viral polymerase gene. The drug can be taken orally and crosses the blood–brain barrier. Most is excreted unchanged in the urine; toxicity is rare.

Virus-infected cells phosphorylate ganciclovir to a monophosphate form, which is then metabolized further to a triphosphate form. The drug is active against herpes simplex virus and cytomegalovirus. It is indicated in the treatment of life- or sight-threatening cytomegalovirus infections in immunocompromised individuals. Ganciclovir causes bone marrow toxicity (monitoring of haematological indices is required during therapy).

Ribavirin is a guanosine analogue that has activity against respiratory syncytial virus, influenza A and B, parainfluenza virus, Lassa fever and other arenaviruses. It inhibits several steps in viral replication including capping and elongation of viral mRNA: its mechanism of action is probably by inhibition of cellular pathways. Usually, ribavirin is administered by aerosol for treatment of severe respiratory syncytial virus infection in infants. It is used in treatment of Lassa fever virus and hantavirus infections.

Anti-retroviral compounds

The introduction of highly active anti-retroviral therapy (HAART) brings about improvement in the CD4 count and a fall in the HIV viral load, with a fall in the incidence of opportunistic infections. Survival and quality of life are improved.

Nucleoside reverse transcriptase inhibitors

Nucleoside reverse transcriptase inhibitors (NRTI) are nucleoside analogues that inhibit the action of reverse transcriptase, the enzyme responsible for the conversion of viral RNA into a DNA copy. These include: the longest established antiretroviral drug zidovudine (AZT), lamivudine (3TC), stavudine (d4T), didanosine (ddI) and zalcitabine (ddC). They are the mainstay of retroviral therapy and are used in combination during initial therapy.

Protease inhibitors

Protease inhibitors are the most effective antiretroviral compounds available because when used as a single agent they produce the greatest fall in viral load. They include indinavir, ritonavir, saquinavir and nelfinavir.

Non-nucleoside reverse transcriptase inhibitors

Non-nucleoside reverse transcriptase inhibitors (NNRTIs) inhibit reverse transcriptase by an alternative mechanism to NRTIs (nevirapine, delavirdine). They have been shown to be effective agents in combination regimens. As resistance occurs after a single mutation, they are only used in maximally suppressive regimens.

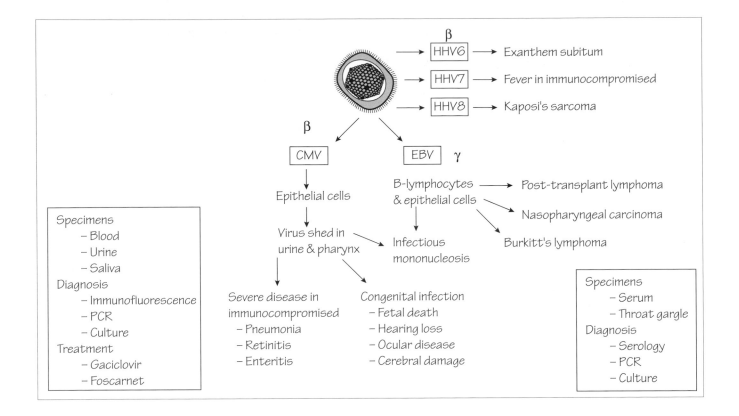

Diagram labels:

β
- HHV6 → Exanthem subitum
- HHV7 → Fever in immunocompromised
- HHV8 → Kaposi's sarcoma

β
- CMV → Epithelial cells → Virus shed in urine & pharynx

EBV γ
- B-lymphocytes & epithelial cells → Post-transplant lymphoma
- → Nasopharyngeal carcinoma
- → Burkitt's lymphoma
- → Infectious mononucleosis

Congenital infection
- Fetal death
- Hearing loss
- Ocular disease
- Cerebral damage

Severe disease in immunocompromised
- Pneumonia
- Retinitis
- Enteritis

Specimens
- Blood
- Urine
- Saliva

Diagnosis
- Immunofluorescence
- PCR
- Culture

Treatment
- Gaciclovir
- Foscarnet

Specimens
- Serum
- Throat gargle

Diagnosis
- Serology
- PCR
- Culture

Herpesviruses

Herpesviruses are large enveloped double-stranded DNA viruses (120–200 nm). The genome is 120–240 kb, coding for more than 35 proteins. The envelope contains glycoproteins and Fc receptors. All infections are lifelong, with latency following the acute primary episode and relapses occurring later in life, especially if the individual becomes immuno-suppressed.

Classification

Herpesviruses are divided into α-herpesviruses (fast-growing cytolytic viruses that establish latent infections in neurones, e.g. herpes simplex and varicella-zoster); β-herpesviruses (slow-growing viruses that become latent in secretory glands and kidneys, e.g. cytomegalovirus); and γ-herpesviruses (latent in lymphoid tissues, e.g. Epstein–Barr virus).

The recently discovered human herpesvirus types 6 and 7 might be classified as γ-herpesviruses because they become latent in lymphocytes, but have most genetic homology with β-herpesviruses and so are classified with this group. The Kaposi's sarcoma-associated virus, human herpesvirus type 8, has homology with Epstein–Barr virus. Herpesvi-ruses are antigenically diverse; only herpes simplex (HSV) 1 and 2 share antigenic similarity.

Cytomegalovirus

Cytomegalovirus (CMV) has a similar structure to other herpesviruses. Infection persists for life and is shed in the urine and saliva. Approximately 50% of adults in the UK have been infected.

Epidemiology and pathogenesis

Infection is transmitted vertically or from close person-to-person contact. Under poor socioeconomic conditions, infection is acquired early but with increasing wealth this is delayed. Pregnant women who develop infection may transmit it to the fetus before or after birth. Infection can also be acquired from blood transfusion or organ transplantation.

Clinical features

In congenital infection neonates may either be severely affected (see Chapter 38), or may be initially asymptomatic, later developing hearing defects or mental retardation. Postnatal infection is usually mild. Immunocompromised patients, especially those with organ transplantation or HIV, may develop severe pneumonitis, retinitis or gut infection either through reactivation of latent virus or acquisition from the donor organ.

Diagnosis
Congenital infection is confirmed by detecting virus in the urine within 3 weeks of birth. In adult infection, CMV can be cultured, or the DNA amplified by PCR, from specimens of urine or blood.

Treatment and prevention
Severe life- or sight-threatening infection should be treated with ganciclovir, together with immunoglobulin in the case of pneumonitis. Appropriate screening of donor organs and blood products can reduce the risk of transmission.

Epstein–Barr virus (EBV)
The virus genome codes for Epstein–Barr nuclear antigen complex (EBNA), latent membrane protein, terminal protein, the membrane antigen complex, the early antigen (EA) complex and the viral capsid antigen.

Epidemiology and pathogenesis
As with CMV, infection is generally found in the very young in developing countries and adults in industrialized countries. Gaining entry via the pharynx, the virus infects B cells and disseminates widely. Epstein–Barr virus is capable of immortalization of B cells. This can result in neoplasia: Burkitt's lymphoma (found in sub-Saharan Africa in association with malaria), nasopharyngeal carcinoma (China), and lymphoma in immunosuppressed patients.

Clinical features
Infection is characterized by fever, malaise, fatigue, sore throat, lymphadenopathy and occasionally hepatitis. It usually lasts about 2 weeks, but persistent symptoms may develop in a few patients. Epstein–Barr virus infection is associated with tumours (see above).

Diagnosis
Diagnosis is made by a rapid slide agglutination technique. Definitive diagnosis is by detection of specific IgM to EBV viral capsid antigen.

Human herpesviruses 6, 7 and 8
These viruses were first isolated in the 1980s. Herpesvirus 6 and 7 have been associated with febrile illness in children, including roseola infantum. These viruses may reactivate in immunocompromised patients but a definite link with symptoms is not proved.

Human herpesvirus 8 has been detected in patients with Kaposi's sarcoma and is the probable cause of this disease.

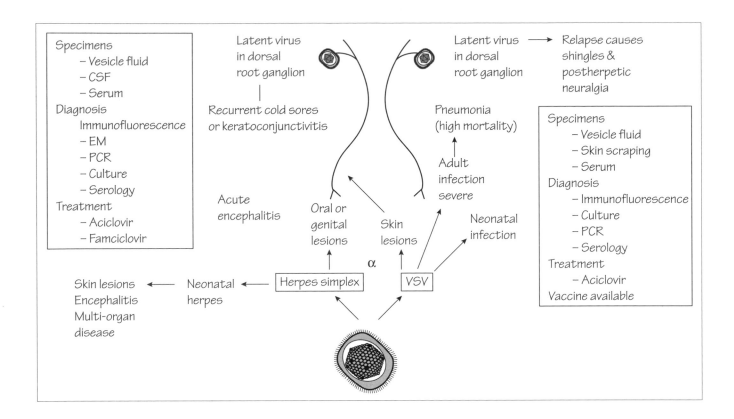

Herpes simplex

Pathogenesis and epidemiology

Herpes simplex is transmitted from person-to-person by direct contact. It causes vesicle formation because of its cytolytic activity. The virus may invade locally and, via the sensory neurones, remains latent in the sensory ganglia. Reactivation may be triggered by physical factors, e.g. infection or sunlight, or psychological stress. As cell-mediated immunity controls infection, patients treated with cytotoxic drugs or having other immunosuppressive conditions are at risk of reactivation and severe infection.

Clinical features

HSV 1 infection is often asymptomatic, but young children commonly develop fever, vesicular gingivostomatitis and lymphadenopathy; adults may exhibit pharyngitis and tonsillitis. Primary eye infection produces severe keratoconjunctivitis: recurrent infection may result in corneal scarring. Primary skin infection (herpetic whitlow) usually occurs in traumatized skin, e.g. fingers. Fatal encephalitis may occur (see Chapter 42). Maternal transmission during childbirth may result in generalized neonatal infection and encephalitis.

HSV 2 infections cause painful genital ulceration that can be severe with symptoms lasting up to 3 weeks. Recurrent infections are milder and virus shedding short-lived but infection can be transmitted to sexual partners during this time. Genital herpes is an important cofactor in the transmission of HIV.

Diagnosis

Under electron microscopy, vesicle fluid may show herpesvirus. Immunofluorescence can also be used to detect HSV. HSV is readily grown in tissue culture from fresh specimens, e.g. vesicle fluid, genital and mouth swabs. Culture from cerebrospinal fluid is often unsuccessful. Polymerase chain reaction detection of virus in cerebrospinal fluid is now the standard method.

The role of serology is limited. Serum:cerebrospinal fluid antibody ratios may help in the diagnosis of HSV encephalitis.

Treatment

Topical, oral and intravenous preparations of antiviral agents, e.g. aciclovir, are available for the treatment of HSV infections.

Varicella-zoster virus

Varicella-zoster virus (VZV) (125 kb) has only one serological type and causes the acute primary infection known as chickenpox or varicella, and its recurrence (shingles).

Pathogenesis and epidemiology

Vesicular fluid contains large numbers of VZV: when the vesicle ruptures, VZV is transmitted by airborne spread. The attack rate in non-immune individuals is very high (>90%).

The incubation period is 14–21 days and the disease is most common in children aged 4–10 years. Individuals are infectious for a few days before the rash appears, until the vesicle fluid has dispersed. Recovery provides lifelong immunity.

The virus lies latent in the posterior root ganglion and in 20% of those previously infected, will travel down the axon to produce lesions in that dermatome known as 'shingles'. As lesions of shingles contain VZV, those who are non-immune may contract chickenpox: it is impossible to contract shingles directly from chickenpox.

Clinical features

Most of the discomfort of VZV infection arises as a result of the rash: systemic symptoms are mild. Individual lesions progress through macules and papules to vesicular eruptions which, following rupture, develop a crust and spontaneously heal. They appear in crops, usually 2 or 3 days apart and affect all parts of the body including the oropharynx and genitourinary tract. The rash lasts for 7–10 days, but complete resolution may take as long again. Rarely, infection can be complicated by haemorrhagic skin lesions that can be life-threatening.

VZV virus pneumonia is more common in adults, especially immunocompromised individuals. It has a high mortality; survivors may recover completely or may have respiratory impairment. Although minor postinfectious encephalitis can also develop, fatal disease is rare. Maternal transmission, through contact with vaginal lesions during birth, can result in severe neonatal infection.

Shingles is a painful condition which is more common in the elderly. Ocular damage may follow the involvement of the ophthalmic division of the trigeminal nerve. Up to 10% of shingles episodes will be followed by postherpetic neuralgia, a condition which may last for many years and is associated with a significant risk of suicide.

Diagnosis

The diagnosis is usually made clinically. Direct staining of vesicle fluid may reveal characteristic giant cells. VZV may be seen by electron microscopy and can be cultivated. Serology is used to determine the immune status of patients considered at risk, e.g. immunocompromised or pregnant. Polymerase chain reaction diagnosis of VZV in cerebrospinal fluid is now a routine technique.

Treatment and prevention

Antiviral agents, e.g. aciclovir, may be used for both adult chickenpox and shingles. The incidence of postherpetic neuralgia may be reduced by the use of amantadine. The pain may be severe and require referral to a pain clinic.

Primary infection can be prevented by a live attenuated virus vaccine but this is not currently part of the routine immunization schedule. Zoster immune globulin can be administered to neonates to prevent serious disease.

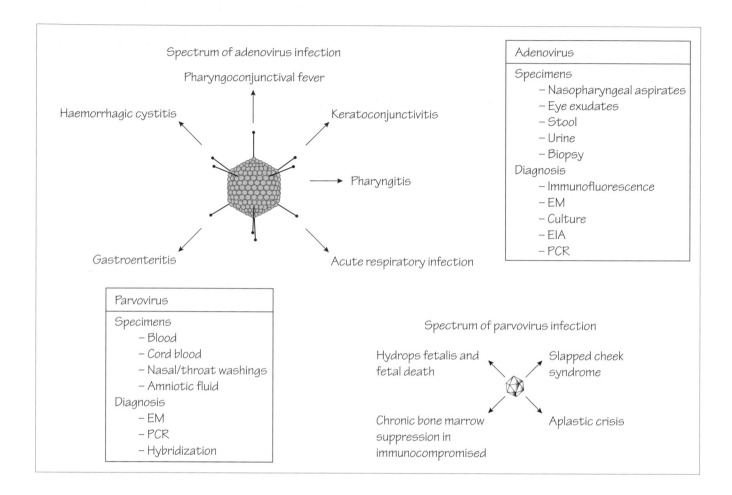

Adenovirus

Adenoviruses are unenveloped, icosohedral, double-stranded DNA viruses. The capsid is made up of 240 hexon and 12 vertex penton capsomeres with vertex fibre projections. The DNA has a viral protein covalently linked to the 5′ end; attachment of this protein is important in infectivity. The hexons possess a group-specific antigen and all human adenoviruses share a common hexon antigen. The pentons also share a group-specific antigen, but the fibres contain type-specific antigens. There are more than 41 serotypes of human adenoviruses divided into six groups (A–F) based on genomic homology.

Epidemiology and pathogenesis

Group A strains cause asymptomatic enteric infection, groups B and C respiratory disease, group D ketatoconjunctivitis, group E conjunctivitis and respiratory disease, and group F infantile diarrhoea. Infection shows little seasonal variation and is usually transmitted by the faecal–oral or respiratory routes. Eye infections, however, can be transmitted by hand–eye contact, a process facilitated in swimming pools; epidemic keratoconjunctivitis is highly infectious.

Clinical features

Clinical syndromes include acute febrile pharyngitis (Serotype1–7), pharyngoconjunctival fever (3, 7) acute respiratory infection and pneumonia (1, 2, 3, 7), conjunctivitis (3, 7), epidemic keratoconjunctivitis (8, 19, 37), enteritis (40, 41) and acute haemorrhagic cystitis (11, 21). Immunocompromised patients may suffer severe pneumonia (1–7), urethritis (37) and hepatitis in liver allografts.

Diagnosis

Virus can be isolated in tissue culture from stool specimens, rectal, conjunctival or throat swabs. Rounding and ballooning indicates infected cells that can be confirmed by specific immunofluorescence. Serotype is determined by using specific neutralizing antibody or haemagglutination inhibition. Enteric adenovirus can be visualized by electron microscopy or detected by EIA or PCR. A four-fold rise in complement fixing group-specific antibody indicates recent infection.

Prevention and control
Live attenuated vaccines are available for types 4 and 7 and are used by military recruits. Swimming pool outbreaks of ocular infection are prevented by adequate chlorination. Likewise, adequate decontamination of equipment and appropriate hygiene by health care staff will prevent transmission between patients undergoing ophthalmic examination.

Parvovirus
Parvoviruses are small, unenveloped, icosohedral, single-stranded DNA viruses (5.6 kb, 8–26 nm). There is only one parvovirus, B19, which is known to cause human disease.

Epidemiology and pathogenesis
Parvovirus B19 must replicate in dividing cells; it targets immature erythrocytes. Clinically, this may produce erythema infectiosum (fifth disease or slapped cheek syndrome), typically in young children. Aplastic crises in patients with chronic haemolytic anaemia, and chronic bone marrow suppression in immunodeficient individuals, may also occur. Rarely, maternal infection may lead to hydrops fetalis and fetal death. Transmission is by the respiratory route; outbreaks of erythema infectiosum may occur in schools. Seroprevalence increases with age with more than 60% of adults possessing antibody.

Clinical features
Erythema infectiosum is a mild febrile disease; children may exhibit a 'slapped cheek' appearance. A symmetrical small joint arthritis may develop, especially in adults. A worsening of anaemia may be the only sign of infection in patients with conditions producing a high red blood cell turnover. The risk of infection in pregnancy is low. There is no evidence that parvovirus causes congenital abnormalities but infection during the second trimester results in spontaneous abortion in 10% of pregnancies.

Diagnosis
Blood, nasal or throat washings, cord blood and amniotic fluid can be examined by electron microscopy. Culture is not used: PCR is the test of choice. Serological testing can be performed.

Prevention and control
No vaccine is available at present. Respiratory precautions should prevent transmission in the hospital environment.

Poxvirus
Historical note
Smallpox was once a major cause of death worldwide. The World Health Organization coordinated an international campaign of vaccination which resulted in the eradication of disease in 1977.

Poxviruses are double-stranded DNA viruses with complex symmetry and a shape resembling a ball of wool. The genome encodes more than 100 proteins.

Monkeypox
This zoonotic organism is found in remote villages in the rainforest of central Africa. The clinical features are similar to smallpox. The disease is severe and the fatality rate is >10% in unvaccinated subjects. Transmission between humans does not occur readily.

Orf
This organism causes cutaneous pustular dermatitis in sheep and goats, which is transmitted to individuals. Human infection is characterized by a single vesicular lesion, typically on the finger, hand or forearm, which resolves spontaneously after a few weeks. Although the virus can be cultivated, diagnosis is usually made by clinical appearance and a history of exposure.

Molluscum contagiosum
This organism causes crops of small regular papular skin lesions, usually on the face, arms, buttocks and back. The pearl-like surface may exhibit a central dimple. Self-inoculation may occur onto damaged skin, e.g. eczema, which may be fostered by the use of topical steroids. It is transmitted by direct contact or via fomites. It is a common condition, especially in children. It may also be sexually transmitted.

Microscopically, the papules are caused by epidermal hypertrophy which extend into the dermis. Cells with inclusion bodies are seen in the prickle-cell layer. The diagnosis is usually made clinically and can be confirmed by electron microscope examination of scrapings from the lesion.

Lesions may continue to appear for up to 1 year in immunocompetent individuals. Molluscum may be a chronic problem for HIV-positive individuals. Traditional treatment—by prodding the lesions with a sharp implement—promotes healing.

Tanapox
Tanapox is a febrile illness with skin lesions that do not ulcerate and heal spontaneously. Infection is acquired in central and east Africa; the diagnosis is usually suggested by travel history.

27 Measles, mumps and rubella

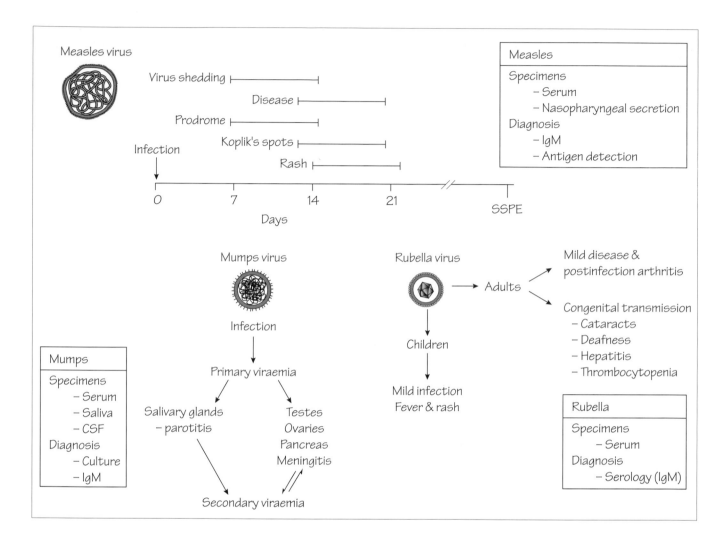

Measles

Measles, a negative single-stranded enveloped RNA virus, is a member of the morbilliform genus of the Paramyxoviridae family. There is a single serotype. The virus encodes six structural proteins including two transmembrane glycoproteins, fusion (F) and haemagglutinin (H), that facilitate attachment to the host cell and viral entry. Antibodies to F and H are protective.

Pathogenesis and epidemiology

Measles virus initially infects the epithelial cells of the upper respiratory tract. Invasion of neighbouring lymphoid tissue leads to primary viraemia and the involvement of the reticuloendothelial system. Invasion of these cells leads to secondary viraemia and dissemination throughout the body, coincident with clinical symptoms.

Measles is transmitted by the aerosol route, with a 95% attack rate in susceptible individuals. The incubation period is 9–12 days. Natural infection is followed by lifelong immunity. In developed countries, mortality is rare except in the immunocompromised, but in developing countries mortality rates may rise to 20%. Mortality is highest in children under 2 years of age and in adults. It has been estimated that a population of 500 000 is required to allow endemic spread. Thus, measles virus dies out naturally in isolated communities, but when the virus is reintroduced attack rates approach 100% and mortality is high.

Clinical features

Infection begins with a 2–4 day coryzal illness, when small white papules (Koplik's spots) are found on the buccal mucosa near the first premolars. The morbilliform rash appearing first behind the ears, spreads centrifugally. After 3–4 days it changes to a brownish colour often accompanied by desquamation. Measles can be complicated by a giant cell pneumonia, secondary bacterial pneumonia and otitis media. Acute postinfectious encephalitis occurs in approximately 1 in 1000 and is associated with a high mortality. Subacute measles encephalitis is a chronic progressive disease occurring mainly in children with leukaemia. It leads to death within a few months. Subacute sclerosing panencephalitis (SSPE) is a rare progressive fatal encephalitis that develops more than 6 years after infection.

Diagnosis

The diagnosis is made clinically and confirmed by serology. Infection in vaccine recipients may be atypical. Primary isolation is difficult and requires primary monkey kidney cells. Virus grows with a giant cell and spindle-shaped cytopathic effect. Serological diagnosis is with haemagglutination inhibition in acute and convalescent serum or an IgM-specific EIA. Salivary IgM detection has also proved valuable. SSPE is diagnosed by detecting virus-specific antibody synthesis in cerebrospinal fluid, e.g. specific IgM.

Ribavirin has activity against measles virus *in vitro* and may be beneficial although no clinical data are available.

Mumps

A member of the *Paramyxovirus* genus, the mumps virus is a pleomorphic, enveloped, antisense RNA virus. It encodes three nucleocapsid proteins: nucleoprotein (NP) and the minor L and P proteins, which participate in RNA replication. There are two envelope glycoproteins, the fusion (F) and haemagglutinin–neuraminidase (HN) and a further membrane associated matrix (M) protein. Mumps has one serotype.

Epidemiology

Infection usually occurs in childhood but many adults are susceptible as it has a lower attack rate than other childhood exanthemas. The incubation period is 14–18 days. Subclinical infection is common, especially in children. Infection is transmitted by droplet spread or direct contact, the virus gaining entry via the upper respiratory tract from where it spreads to local lymph nodes. The virus invades the salivary glands, testes, ovaries, central nervous system and pancreas. Natural infection is followed by lifelong immunity.

Clinical features

Mumps is characterized by fever, malaise, myalgia and inflammation of the parotid glands. Meningitis occurs in up to 10% of patients with parotitis: mumps virus was once one of the most common causes of viral meningitis. Complete recovery is almost invariable, although rare fatal forms and postmeningitis deafness may occur. Other complications, including orchitis, oophoritis or pancreatitis, are more common in adolescents and young adults, often after the parotitis has resolved.

Diagnosis

Typical mumps does not usually require laboratory diagnosis. Mumps virus can be isolated from saliva. Serological diagnosis is made by showing a rise in acute and convalescent titres or the presence of specific IgM. Reverse transciptase PCR is available.

Rubella

Rubella, a rubivirus and member of the Togaviridae, is an isocohedral, pleomorphic, enveloped, positive-strand RNA virus. The virus produces three structural proteins: two envelope glycoproteins, E1 and E2, and a nucleocapsid protein C. There is only one antigenic type of rubella virus.

Epidemiology

Peak incidence occurs in late winter and early spring. Transmission is via the respiratory route. Virus is shed from 7 days before and 14 days after the appearance of the rash. Maternal infection during the first trimester may cause congenital abnormalities in approximately 60% of cases. The risk to the fetus is greatest during the first trimester. Natural infection is followed by solid immunity.

Clinical features

Rubella is associated with fever, a fine red maculopapular rash and lymphadenopathy. During the prodrome red pin-point lesions occur on the soft palate. Arthritis (more common in females) and self-limiting encephalitis are complications.

Congenital infection is associated with fetal death or severe abnormalities, e.g. deafness, central nervous system deficit, cataract, neonatal purpura and cardiac defects.

Diagnosis

Acute diagnosis in adults is sometimes required in pregnancy or pregnancy contact. Serological screening for the presence of rubella antibody is performed by radial haemolysis, agglutination techniques, EIA and haemagglutination inhibition. Congenital disease is confirmed by demonstrating the presence of specific IgM persistent antibodies (>6 months) in an infant, or viral isolation from the infant after birth.

Prevention of measles, mumps and rubella

These three diseases can be prevented by a live attenuated combined vaccine (MMR). Vaccine is given between 13 and 15 months, with a booster dose given at school entry. Booster doses of measles vaccine may be required. The rapid antibody response to measles vaccine can be used to protect susceptible individuals exposed to measles. Women who are sexually active should be screened for the presence of rubella antibodies, e.g. when they first attend for contraceptive advice. Seronegative women should be offered rubella vaccination.

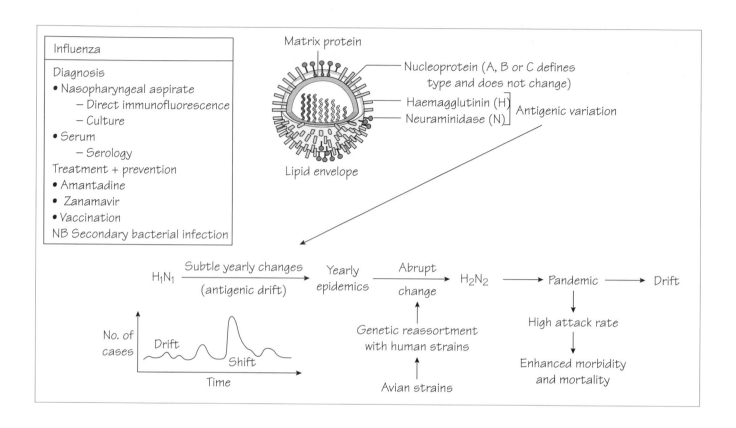

Influenza virus

Influenza virus, an orthomyxovirus, is an enveloped spherical (100 nm) virus. It contains a negative single-stranded RNA genome divided into eight segments. This facilitates genetic reassortment allowing the development of different viral surface antigens. The virus expresses seven proteins, three of which are responsible for RNA transcription. The nucleoprotein has three antigenic types designating the three main virus groups, influenza A, B and C. Of the three, influenza A, and rarely B, undergo genetic shift. The matrix protein forms a shell under the lipid envelope with haemagglutinin and neuraminidase proteins expressed as 10 nm spikes on the envelope. These interact with host cells. Virus is typed based on the haemagglutinin (H) and neuraminidase (N) antigens.

Epidemiology

Annual epidemics of respiratory infection occur because of minor antigenic changes (antigenic drift). When there is a major antigenic shift, a worldwide pandemic may develop. Pandemics occur every 10–40 years, often originating in the Far East and carried by global atmospheric circulation in a westerly direction. Such strains can often be traced to infected poultry or pigs. Influenza A circulates in animal populations and is more likely to cause a pandemic; B and C are exclusively human pathogens. Pandemic strains have a high attack rate and are associated with increased morbidity and mortality, e.g. 20 million died in the 'Spanish flu' epidemic of 1919.

Clinical features

There is a 1–4 day incubation period, with patients infectious for a day preceding and the first 3 days of symptoms. Headache, myalgia and fever last for 3–4 days and are often associated with a cough that persists for a few days longer. Complications, which are more common in the elderly and patients with cardiopulmonary disease, include primary viral or secondary bacterial pneumonia. In children, the use of salicylates in the treatment of influenza may give rise to Reye's syndrome.

Diagnosis

Laboratory diagnosis for individual cases is rarely required but is important for surveillance and the prevention of epidemic disease. Virus can be isolated in embryonated hens' eggs or continuous canine kidney cells or monkey kidney. Rapid identification of virus type is needed to plan the composition of influenza vaccine, a process that is coordinated internationally. Serology provides a retrospective diagnosis. RT-PCR is available.

Treatment, prevention and control

Influenza is treated symptomatically; secondary bacterial infections require appropriate antibiotics. Inactivated viral vaccines are prepared each year, reflecting the currently circulating viruses. This provides 70% protection and is recommended for individuals at risk of severe disease, e.g. cardiopulmonary disease and asthma. Influenza A infection can be prevented or modified in susceptible individuals with amantadine or rimantadine. New neuraminidase inhibitors, such as zanamavir, demonstrate good activity against influenza A and B.

Parainfluenza virus

A fragile enveloped paramyxovirus (150–300 nm) containing a single strand of negative-sense RNA (15 kb), it has four types that share antigenic determinants. The lipid envelope expresses two glycoprotein 10 nm 'spikes': the haemagglutinin–neuraminidase (HN) and the fusion protein (F).

Pathogenesis and epidemiology

The virus attaches to host cells via the HN molecule; the envelope fuses with the host cell via the F molecule. The virus multiplies throughout the tracheobronchial tree. Infection, transmitted by the respiratory route, peaks in the winter with the highest attack rates in children under 3 years.

Clinical features

In this common self-limiting condition, usually lasting 4–5 days, children are distressed, coryzal and febrile. In young children, hoarse coughing often alternates with hoarse crying and is associated with inspiratory stridor secondary to laryngeal obstruction (croup). Rarely, bronchiolitis, bronchopneumonia or acute epiglottitis may develop signalled by reduced air entry and cyanosis.

Diagnosis and treatment

As most cases are treated in the community, diagnosis is primarily clinical. Direct immunofluorescence gives rapid results. Viral isolation is possible and RT-PCR is available.

Most infections are managed at home with symptomatic relief, e.g. paracetamol and humidification. Severe infection can be treated with ribavirin and humidified oxygen.

Respiratory syncytial virus

An enveloped paramyxovirus (120–300 nm) containing a single strand of negative-sense RNA, it attaches to host cells by glycoprotein antigen, which appear as 12 nm projecting spikes on the lipid envelope. There is antigenic variation within the two types, A and B.

Epidemiology

Respiratory syncytial virus (RSV) is found world-wide, infecting children during the first 3 years of life. There are yearly epidemics in the winter months in temperate countries or in the rainy season in tropical countries. RSV spreads readily in the hospital environment.

Clinical features

Following an incubation period of 4–5 days, children may develop coryza. The lower respiratory tract is involved in approximately 40% causing bronchitis in older children and bronchiolitis in the very young. Severe disease may develop quickly but, with intensive care, mortality is very low. Children with bronchiolitis are febrile, have a rapid respiratory rate, hyperinflation, with wheezing and crepitations that may change on repeated examination. Cyanosis is rare. Radiological appearances are variable but signs of hyperinflation and increased peribronchial markings are common.

Diagnosis and treatment

Rapid laboratory diagnosis may be obtained by direct immunofluorescence or EIA of nasopharyngeal secretions. There is little role for serological diagnosis. Virus can be cultivated in HeLa and Hep-2 cells. RT-PCR is available.

The treatment for RSV infection is based on symptomatic relief and steam. More severe cases may require hospitalization and humidified oxygen. Severely ill patients may benefit from aerosolized ribavirin.

Prevention

Previous attempts to produce a killed vaccine made vaccinated infants more susceptible to severe disease. No vaccine is currently available.

Coronavirus

A spherical enveloped virus (80–160 nm) containing a single strand of positive-sense linear RNA (27 kb), the envelope contains widely spaced club-shaped spikes. Coronaviruses cause a coryza-like illness similar to that of rhinovirus. The virus has been observed in the faeces of patients with diarrhoeal disease and asymptomatic subjects. Diagnosis is by serology using CF or EIA, detection of coronavirus-specific antigens or by electron microscopy.

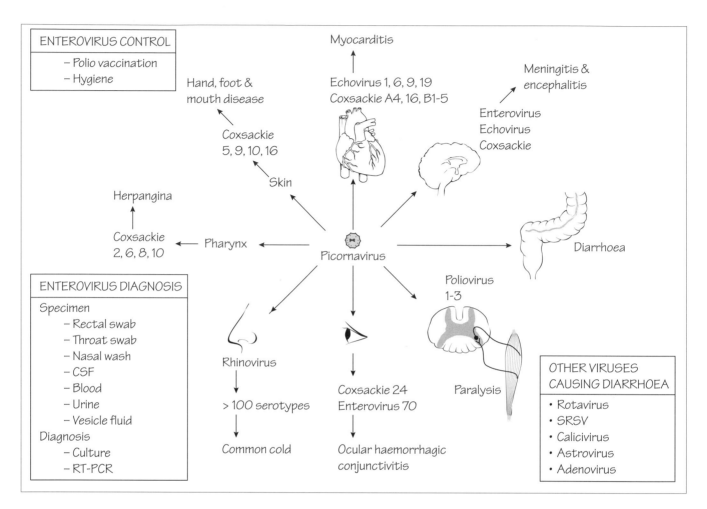

ENTEROVIRUS CONTROL
– Polio vaccination
– Hygiene

Myocarditis

Echovirus 1, 6, 9, 19
Coxsackie A4, 16, B1-5

Meningitis & encephalitis

Enterovirus
Echovirus
Coxsackie

Hand, foot & mouth disease

Coxsackie 5, 9, 10, 16

Skin

Herpangina

Coxsackie 2, 6, 8, 10 ← Pharynx

Picornavirus

Diarrhoea

Poliovirus 1-3

ENTEROVIRUS DIAGNOSIS

Specimen
– Rectal swab
– Throat swab
– Nasal wash
– CSF
– Blood
– Urine
– Vesicle fluid
Diagnosis
– Culture
– RT-PCR

Rhinovirus

> 100 serotypes

Common cold

Coxsackie 24
Enterovirus 70

Ocular haemorrhagic conjunctivitis

Paralysis

OTHER VIRUSES CAUSING DIARRHOEA
• Rotavirus
• SRSV
• Calicivirus
• Astrovirus
• Adenovirus

Enterovirus

Enteroviruses and rhinoviruses belong to Picornaviridae. Enteroviruses fall into three main serotype groups: poliovirus, coxsackie virus and enteric cytopathic human orphan (ECHO) virus. Later viral serotypes have been given a number (enterovirus 68–72).

The viruses are unenveloped, isocohedral, symmetrical particles with single-strand positive RNA coding for four proteins, VP1–4. Antibodies derived from intact virus are type specific but those derived from empty capsids cross-react with other enteroviruses.

Pathogenesis

Enteroviruses attach and enter cells by a specific receptor that differs for different virus types; receptor differences may influence tissue tropism. Enteroviruses usually enter the body by the intestinal tract. This is followed by a local viraemia and invasion of susceptible reticuloendothelial cells. A subsequent major viraemia leads to invasion of target organs, e.g. meninges, spinal cord, brain or myocardium. The poliovirus appears to spread along nerve fibres. If significant multiplication occurs within the dorsal root ganglia the nerve fibre may die, with resultant motor paralysis.

Epidemiology

Enteroviruses spread by the faecal–oral route. In developing countries infection occurs early in life: it occurs later in industrialized countries. Cases of polio have been documented in parents and carers of infants receiving live vaccine. Adult non-immunity may be because of the absence of a polio vaccination programme, e.g. immigrant populations. Those at risk should receive immunization at the same time as the infant. In the tropics infection occurs throughout the year but in temperate countries it peaks during the summer.

Clinical features

Polio may present as a minor illness (abortive polio), aseptic meningitis (non-paralytic polio), paralytic polio with lower motor neurone damage and paralysis, or progressive postpoliomyelitis muscle atrophy (a late recrudescence of muscle wasting sometimes decades after the initial paralytic polio). In paralytic polio muscle involvement is maximal within a few days after commencement of paralysis: recovery may occur within 6 months.

Self-limiting aseptic meningitis is a common presentation of enterovirus infection (see Chapter 42), although severe focal encephalitis or general infection may present in neonates. Herpangina is a self-limiting painful vesicular infection of the pharynx caused by some coxsackievirus types. Coxsackie B viruses are important causes of acute myocarditis (see Chapter 41). Hand, foot and mouth disease is characterized by a vesicular rash of the palms, mouth and soles that heals without crusting.

Diagnosis and treatment

Most enteroviruses are readily cultivated in tissue culture. In patients with signs of meningitis a specimen of CSF, a throat swab and faecal specimen should be sent for viral culture, although RT-PCR is used increasingly. With a multiplicity of enterovirus types, serological diagnosis is impractical (complement fixation and EIA are available).

Clinical trials of the experimental drug pleconaril show benefit in the treatment of enteroviral meningitis. However, the mainstay of management remains supportive care. Artificial ventilation may be required in the case of polio.

Prevention

Polio can be prevented by vaccination but the efficiency is dependent on an adequate population uptake. Two vaccines are available: the oral live attenuated Sabin, and the killed parenteral Salk vaccine. The Sabin vaccine is used in the international polio eradication campaign. It provides rapid immunity and displaces the wild-type virus circulating in communities. It is part of the routine infant immunization programme and for those at increased risk of exposure, e.g. travellers and some health-care workers. The Sabin vaccine is contraindicated in immunocompromised individuals, where the Salk vaccine is used.

Rhinovirus

The classical cause of the common cold, there are more than 100 serological types. During the short incubation period (2–4 days), the virus infects the upper respiratory tract, invading only mucosa and submucosa. Viral excretion is highest during symptoms: transmission is by close contact with infected individuals. Headache, nasal discharge, upper respiratory tract inflammation and fever may be followed by secondary bacterial infections, e.g. otitis media and sinusitis. Infection occurs world-wide with a peak incidence in the autumn and winter. Immunity after infection is poor because of the multiplicity of serotypes. No specific treatment is available and a vaccine is impractical.

Rotavirus

Rotaviruses are unenveloped spherical viruses in a double capsid shell. They contain 11 double-stranded RNA segments coding for nine structural proteins and several core proteins. Rotaviruses possess group, subgroup and serotype antigens.

Pathogenesis

Rotaviruses infect small intestinal enterocytes: damaged cells are sloughed into the lumen, releasing viruses. Diarrhoea is caused by poor sodium and glucose absorption by the immature cells that replace the damaged enterocytes.

Epidemiology

Rotavirus infection is the most important cause of viral diarrhoea, most commonly in children between 6 months and 2 years of age. Morbidity is highest in the young. There are seasonal peaks in the winter in temperate countries. Antibody to the virus does not confer immunity to further infection.

Diagnosis

Laboratory diagnosis is by electron microscopy or EIA.

Treatment and prevention

Treatment is symptomatic and supportive. The risk of infection can be reduced by provision of adequate sanitation. Live attenuated vaccines are being developed.

Small round structured virus, calicivirus and astrovirus

Small round structured virus (SRSV) is a human calicivirus. It is unenveloped, containing a single strand of positive-sense RNA and has a single capsid protein. Caliciviruses are antigenically distinct from SRSV. Astroviruses are small spherical particles: more than five serotypes have been recognized.

Virus replication occurs in the mucosal epithelium of the small intestine resulting in broadening and flattening of the villi and crypt cell hyperplasia.

These viruses cause a mild self-limiting acute diarrhoeal disease, although SRSV may present with sudden onset projectile vomiting and explosive diarrhoea. Infection is transmitted by the faecal–oral route with symptoms developing after a short incubation period (24–48h). Sudden outbreaks of SRSV infection may occur in institutions including hospitals; respiratory transmission has been postulated but never proved.

Diagnosis is by electron microscopy and immunoassays are being developed for routine use. RT-PCR is available.

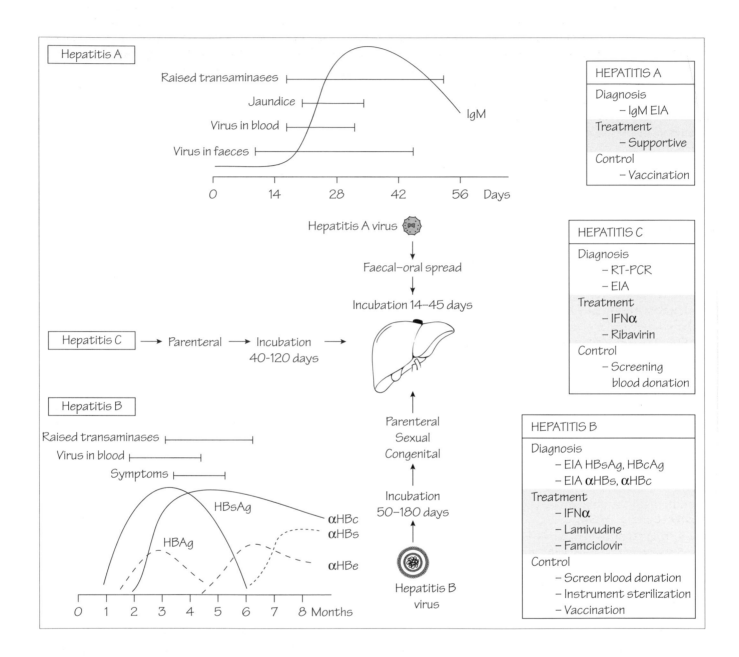

Viral hepatitis must be distinguished from other viral causes of hepatitis, e.g. CMV, herpes simplex, EBV, as well as leptospirosis and non-infective causes.

Hepatitis A

Hepatitis A virus (HAV) is a *Enterovirus* in the family Picornaviridae. It has one serotype. The genome is a single strand of positive-sense RNA.

Transmission is faecal–oral associated with summer, institutional outbreaks and point-source outbreaks following faecal contamination of water or food, e.g. oysters.

Seroprevalence is highest in individuals in lower socio-economic groups. Anicteric infection is more common in the young, the risk of symptomatic disease increasing with age.

Clinical features

Infection is characterized by a flu-like illness followed by jaundice. Most patients make an uneventful recovery.

Diagnosis

Diagnostic antibodies, anti-HAV IgM, appear before jaundice develops and persist for 3 months. IgG antibodies can also be used to determine the immune status.

Treatment and prevention
Symptomatic treatment and support are all that are usually necessary. Chronic hepatitis does not occur.

Adequate sanitation and good personal hygiene will reduce the transmission of HAV. An effective inactivated vaccine is available for active protection. Individuals can be passively protected using human immunoglobulin.

Hepatitis B
Hepatitis B (HBV), a hepadnavirus, is a small enveloped virus containing 3.2 kb partially double stranded DNA which codes for three surface proteins, e.g. surface antigen (HBsAg), core antigen (HBcAg), precore protein, a large active polymerase protein and transactivator protein.

Hepatitis B is transmitted by parenteral, congenital, and sexual routes.

Clinical features
There is a long incubation period (up to 6 months) before acute hepatitis develops, often of insidious onset. Fulminant disease carries a 1–2% mortality. Up to 10% of patients develop a chronic hepatitis complicated by cirrhosis or hepatocellular carcinoma. Congenital infection brings a high risk of hepatocellular carcinoma.

Diagnosis
Immunoassays for HBsAg, HBeAg, HBcAg and associated antibodies enable both the diagnosis of acute infection and previous exposure.

Treatment and prevention
Interferon (IFNα) treatment has limited efficacy but nucleoside analogues, e.g. lamivudine and famciclovir, appear to have value.

Those at high risk should be immunized with recombinant HBV vaccine. Administration of vaccine and specific immunoglobulin to neonates of infected mothers reduces transmission. Blood donations are screened. Needle exchange programmes for drug abusers and sex health education schemes are beneficial.

Hepatitis C
Hepatitis C (HCV) is a positive-stranded RNA virus encoding a single polypeptide.

Infection is mainly transmitted through infected blood. Seroprevalence is <1% in healthy blood donors, higher in developing countries and highest in high-risk groups, e.g. those receiving unscreened transfusions. Health-care workers are at risk. Vertical transmission is rare.

Clinical features
Infection may cause a mild acute hepatitis but many cases are asymptomatic: fulminant disease is rare. HCV infection persists in up to 80% of patients: up to 35% of these develop cirrhosis, liver failure and hepatocellular carcinoma between 10 and 30 years later. This occurs because frequent virus mutation results in immunologically distinct 'quasi-species', allowing the organism to escape immunological control.

Diagnosis
HCV cannot be cultured: diagnostic antibody may be detected by EIA. RT-PCR amplification also provides genotyping and confirms active disease. The efficacy of treatment can be monitored by measuring HCV RNA concentrations.

Treatment and prevention
Ribavirin therapy is associated with a reduction in transaminase levels that often relapses after cessation: with interferon-α, the combination is more effective. Hepatitis C infection is a major indication for liver transplantation.

Similar measures to those employed against HBV will also prevent transmission of HCV. There is no vaccine.

Hepatitis D
This defective RNA virus is surrounded by an HBsAg envelope. It is transmitted by intimate contact or by blood products and produces disease after a short incubation period either as a coinfection with HBV, or a superinfection in an HBV carrier. Although asymptomatic infection may occur, hepatitis D (HDV) is associated with severe hepatitis and an accelerated progression to carcinoma. Detection of the HDV antigen or IgM antibodies to it by EIA confirms the diagnosis. Preventive measures for HBV also protect against HDV.

Hepatitis E
This virus is related to calicivirus and is a small, single strand, non-enveloped RNA virus. Transmitted by the faecal–oral route, outbreaks may occur after contamination of water supplies. The diagnosis is made by detecting specific IgM. Infection is prevented by appropriate hygiene measures.

New agents
Two new viruses, hepatitis F and hepatitis G, have been described recently: their clinical significance is not yet known.

31 Tropical, exotic or arbovirus infections

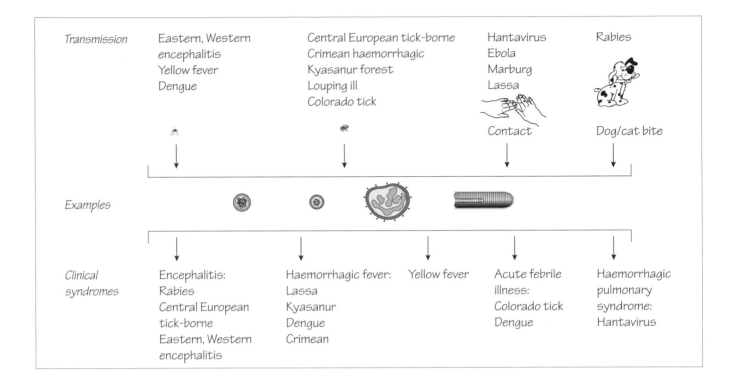

Transmission	Eastern, Western encephalitis Yellow fever Dengue	Central European tick-borne Crimean haemorrhagic Kyasanur forest Louping ill Colorado tick	Hantavirus Ebola Marburg Lassa Contact	Rabies Dog/cat bite
Examples				
Clinical syndromes	Encephalitis: Rabies Central European tick-borne Eastern, Western encephalitis	Haemorrhagic fever: Lassa Kyasanur Dengue Crimean	Yellow fever	Acute febrile illness: Colorado tick Dengue	Haemorrhagic pulmonary syndrome: Hantavirus

More than 100 viral infections cause encephalitis or haemorrhagic fever. Almost all are zoonoses, where the human is an accidental host that has come in contact with the natural life-cycle. They belong to many virus families, e.g. alphavirus, flavivirus, bunyavirus, rhabdovirus, arenavirus and filovirus. They are transmitted by direct contact with blood and body fluids or by the bite of arthropods, such as mosquitoes, ticks and sandflies (these are known as arboviral infections). Infection is associated with a high mortality in many instances.

Rabies
Rabies is an acute rhabdovirus infection that, once symptoms develop, causes a fatal encephalomyelitis.

It is caused by a bullet-shaped RNA rhabdovirus with helical nucleocapsid, enclosed in a membrane with 6–7 nm glycoprotein spikes and a single strand of negative-sense RNA. It infects warm-blooded animals world-wide. The virus is found in the saliva and it is transmitted to humans through the bite of an infected animal. Two epidemiological patterns of disease exist: urban rabies, a disease of feral and domestic dogs; and sylvatic rabies, which occurs in small carnivores in the countryside. Dog-bites are responsible for most infections. Vampire bats are an important reservoir and vector of infection in the

Americas, whereas the red fox is the major reservoir of infection in Europe.

The virus gains entry to the nervous system via the motor end plates and spreads up the axons and spinal cord to enter the brain. Sites with short neural connections to the central nervous system have the shortest incubation period (7 days), whereas a bite on the foot may have an incubation period of 100 days. The depth of the bite and the concentration of viral inoculum also influence the incubation period.

A prodromal fever, nausea and vomiting precede the characteristic features of disease which takes one of two forms: furious (hyperexcitability, hyper-reactivity, hydrophobia) and dumb rabies (an ascending paralysis). Disease is progressive and inevitably fatal. Diagnosis is based on the clinical features or an exposure history. It may be confirmed by specific fluorescence in corneal scrapings or brain biopsy, or by finding specific rabies antibody in serum or tears. Disease may be prevented by appropriate wound care, local antiserum, systemic hyperimmunoglobulin and a postexposure vaccination course. The human diploid cell vaccine is used for pre-exposure as well as post-exposure prophylaxis.

Yellow fever
Yellow fever virus is a flavivirus, an enveloped RNA virus

with a single molecule of single-stranded RNA. It is transmitted to humans by the bite of an *Aedes aegypti* mosquito. Yellow fever is a zoonosis in which humans are an accidental host (sylvatic disease) but an urban cycle results in periodic human epidemics. Infection may be asymptomatic or cause an acute hepatitis and death resulting from necrotic lesions in the liver and kidney. Following an incubation period of 3–6 days, the patient presents with fever, nausea and vomiting. A few days later, the patient becomes toxic and jaundiced with proteinuria. Haemorrhagic manifestations may develop and the vomitus may be black with digested blood (vomito negro). The mortality rate is high but patients who recover, do so completely. Diagnosis is confirmed by viral culture and serology. Disease prevention is by mosquito control and human vaccination with the live attenuated yellow fever vaccine.

Dengue

This mosquito-borne flavivirus is closely related to yellow fever virus. It has four serotypes. Infection is transmitted by *Aedes* mosquitoes; the incubation period is 2–15 days. A viraemia is present at the onset of fever and persists for several days. Dengue virus is found throughout the tropics, e.g. India, Southeast Asia, Pacific islands, Caribbean islands, South America, Africa and the Middle East. Dengue epidemics occur when a new serotype enters the community or a large number of susceptible individuals move into an endemic area. Urban epidemics can be explosive and severe.

Following the sudden onset of fever and chills, headache and malaise, patients complain of pains in the bones and joints. Fever may be biphasic and a mild rash may also be present. Dengue haemorrhagic syndrome is a more severe form of the disease with severe shock and bleeding diathesis. The mortality rate is 5–10%.

Dengue infection may be confirmed by serological means, culture and PCR-based amplification methods. Dengue can only be prevented by controlling mosquito populations.

Treatment of acute cases is symptomatic.

Japanese B encephalitis

This is a mosquito-borne flavivirus infection that causes encephalitis with a high mortality. Pigs are the natural reservoir of the infection. Patients present with abrupt onset fever and severe headache, nausea and vomiting.

Convulsions can occur. There may be permanent cranial nerve or pyramidal tract damage. The infection may be prevented by vaccination. A similar disease pattern occurs with St Louis, Murray Valley, West Nile and Rocio viruses.

Lassa fever

Lassa fever is a severe haemorrhagic fever caused by an arenavirus. Infection is transmitted from the reservoir house rat to humans and from person to person by contact. The virus can affect all organs. Patients may present with fever, mouth ulcers, myalgia and haemorrhagic rash. Diagnosis is based on clinical symptoms and a history of exposure. Laboratory confirmation is by culture, RT-PCR or serology, e.g. immunofluorescence. Ribavirin is effective, especially if given early in the course of disease and can be given as postexposure prophylaxis to close contacts of cases.

Ebola and Marburg virus

These infections are found in Africa transmitted to humans from primates or from the putative rodent reservoir host. They causes an acute haemorrhagic disease with high fever, bleeding, toxicity and shock; the mortality is high. Infection can be transmitted by close contact, especially in the hospital environment. Treatment is supportive and benefit may be obtained from hyperimmune serum. Community control is not possible as the reservoir is not confirmed, but strict blood and body fluid precautions will prevent transmission in hospital.

Hantavirus

This bunyavirus infection is transmitted to humans from rodents and causes a haemorrhagic fever with renal failure. The disease occurs widely throughout the world including North and South America, the Far East, Africa and some Pacific islands. Different hantaviruses have been associated with outbreaks throughout the world. Person-to-person spread does not appear to take place. The incubation period is 2–3 weeks followed by fever, headache, backache and injected conjunctiva and palate. Hypotension, shock and oliguric renal failure follow. The mortality rate is about 5%.

Diagnosis of hantavirus infection is difficult but type-specific serological tests are being developed. Ribavirin appears to improve mortality but controlled clinical trials have not been performed.

32 Yeast infections

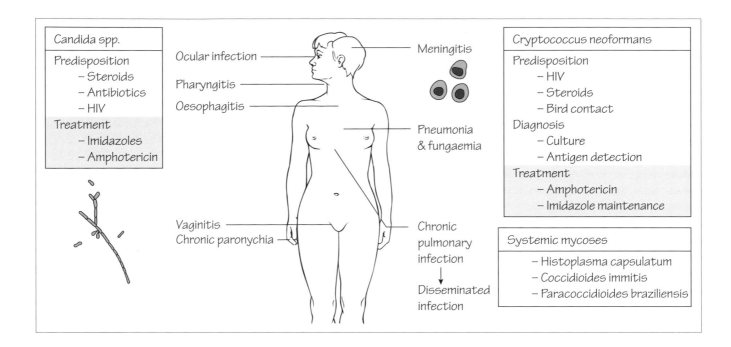

Fungi cause a wide range of diseases ranging from cutaneous dermatophyte infections to invasive infection in the severely immunocompromised patient. They may have a yeast-like morphology, (see below), or be filamentous (see Chapter 33).

Candida spp.

Candida spp. are widely distributed in the environment. They form part of the normal commensal population of the skin, the gastrointestinal tract and genitourinary tract. Following the use of broad-spectrum antibacterials, fungal overgrowth may occur which may develop into infection. Patients with immunodeficiencies are particularly susceptible to this progression. Most infections are caused by *C. albicans*. Infection with other species, e.g. *C. tropicalis*, *C. parapsilosis*, *C. glabrata* and *C. pseudotropicalis*, is becoming more frequent because they may be resistant to the antifungal agents used in therapy or prophylaxis.

Pathogenesis

Although these organisms have adhesins and extracellular lipases and proteinases, they have little capacity to invade. Infection occurs when the natural resistance provided by the normal bacterial flora is altered by antibiotics, or where there is a severe loss in T cell function.

Clinical features

Candida spp. cause pain and itching with creamy curd-like plaques on mucosal surfaces which bleed when removed. Skin and nailbed infections are common. In the immunocompromised patient, pharyngitis and oesophagitis can be severe; the associated dysphagia may lead to weight loss, an AIDS-defining illness. Systemic invasion is common in neutropenic patients.

Laboratory diagnosis

As *Candida* spp. form part of the normal flora, the significance of individual isolates can only be determined in relation to the overall clinical picture. *Candida* may be visualized on a wet preparation or Gram-stained smear. It grows readily on simple laboratory media: Sabouraud's dextrose agar is used if a selective medium is required.

Candida albicans is identified by demonstrating germ tube formation, or characteristic growth on chrome agar. Species identification is by biochemical techniques.

Antifungal susceptibility

Candida spp. are susceptible to amphotericin, with the exception of *C. lusitaniae*. They are usually susceptible to the imidazoles, e.g. fluconazole and to 5-flucytosine.

Cryptococcus neoformans

Cryptococcus neoformans is the only species of this genus that regularly causes infection in humans. It is a saprophyte and animal commensal; pigeon faeces composition favours its growth. It is a rare cause of chronic lymphocytic meningitis in patients with lymphoma, those taking steroid or cytotoxic therapy and those with intense exposure, e.g. pigeon fanciers. *Cryptococcus* is recognized as an important pathogen in AIDS patients as up to 15% become infected.

Pathogenesis
The organism possesses an antiphagocytic capsule and several lytic enzymes, but are dependent on impaired T cell-mediated immunity.

Clinical features
Infection usually presents as subacute meningitis although pneumonia and fungaemic shock are recognized. In AIDS patients, relapses are common and life-long suppressive therapy is necessary.

Laboratory diagnosis
It may be directly visualized in CSF by Gram's staining or India ink. A latex test can detect capsular polysaccharide antigen. The organism may be isolated on blood or Sabouraud's agar; it is identified by biochemical tests.

Treatment
Amphotericin is the treatment of choice; liposomal preparations may be used to reduce toxicity. Fluconazole may also be used.

Pityriasis versicolor
Malassezia furfur infects the stratum corneum causing brown scaly macules. Patients with AIDS may develop severe dermatitis. Topical application of antifungal agents is usually successful.

Systemic yeast infections
Five main species are associated with systemic infection: *Histoplasma capsulatum*, *Histoplasma capsulatum* var. *duboisii*, *Blastomyces dermatitidis*, *Coccidioides immitis* and *Paracoccidioides brasiliensis*.

Infection is acquired by the respiratory route. They have a defined geographical distribution: south-west USA, South America and Africa. Severe disease is more likely in patients with reduced cell-mediated immunity.

Clinical features
Although usually asymptomatic or self-limiting, pulmonary or cutaneous infection may disseminate in infants or the immunocompromised causing severe illness.

Laboratory diagnosis
These organisms are hazardous, and should be handled in a specialized containment facility. Yeast cells are visualized in wet preparations of sputum, CSF, urine, pus or skin scrapings, or with Gram's stain. They may be isolated from blood culture specimens.

Treatment
Patients with severe disease may be treated with amphotericin B.

Azoles
The azole group of compounds (clotrimazole, miconazole, fluconazole and itraconazole) act by blocking the action of cytochrome P450 and sterol 14αdemethylase. This latter enzyme allows the incorporation of 14-methyl sterols in the fungal membrane, instead of ergosterol. Resistance can develop during long-term treatment.

Clotrimazole and miconazole are frequently used as topical preparations for minor infections.

Fluconazole
Fluconazole can be given orally, topically and parenterally, is widely distributed, crosses the blood–brain barrier and is active against *Candida* and *Cryptococcus* but not filamentous fungi. It is used for the prophylaxis and treatment of cryptococcal infections and treatment of superficial and systemic candidiasis. Although well tolerated, it may cause liver enzyme abnormalities. It has significant drug interactions, increasing the serum concentration of phenytoin, cyclosporin and oral hypoglycaemic agents and reducing the rate of warfarin metabolism.

Itraconazole
In addition to being effective against *Candida*, *Cryptococcus neoformans* and *Histoplasma,* itraconazole also displays activity against filamentous fungi, including *Aspergillus* and the dermatophytes. It is indicated in treatment of invasive candidiasis, cryptococcosis, aspergillosis, superficial mycoses and pityriasis versicolor. Resistance is rare.

Highly lipid soluble, it is incompletely absorbed from the gastrointestinal tract; optimal absorption is achieved when taken with food. The drug achieves high tissue concentrations.

Adverse events are rare but there may be a transient increase in the transaminases and in the concentrations of cyclosporin, digoxin and phenytoin. Concomitant use of rifampicin or rifabutin lowers the serum concentration of itraconazole.

Flucytosine
This synthetic fluorinated pyrimidine inhibits *Candida* spp., *Cryptococcus neoformans* and some moulds. The drug disrupts protein synthesis. It is well absorbed orally and can be given intravenously. Bone marrow suppression, thrombocytopenia and abnormal liver function tests. Resistance develops rapidly with monotherapy.

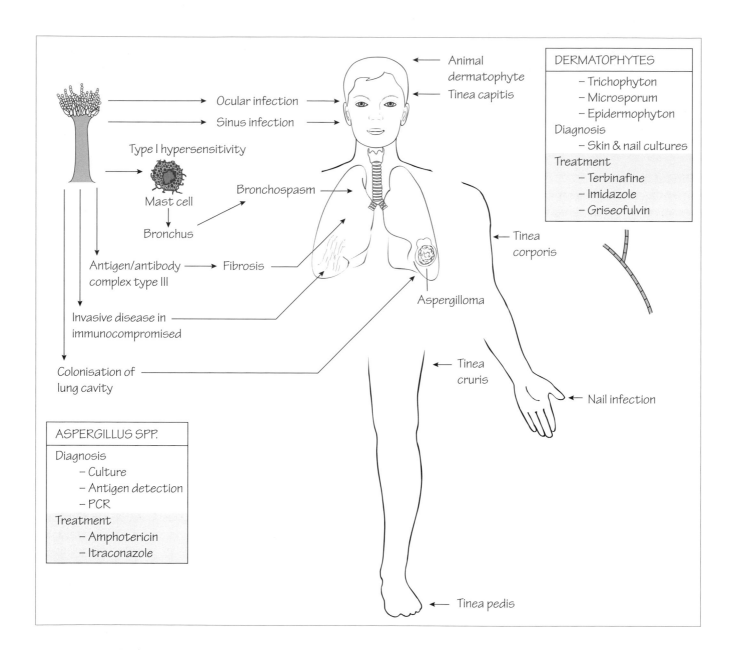

Aspergillus spp.

Aspergillus spp. are ubiquitous, free-living, saprophytic organisms. Four are regularly associated with human infection: *A. fumigatus*, *A. niger*, *A. flavus* and *A. terreus*.

Clinical features

Inhalation of *Aspergillus* spores may give rise to a type III hypersensitivity reaction with fever, dyspnoea and progressive lung fibrosis (farmer's lung). Some patients colonized by *Aspergillus* develop a type I hypersensitivity reaction resulting in intermittent airway obstruction (bronchopulmonary allergic aspergillosis).

Healed cavities or old bronchiectasis can become colonized with *Aspergillus*, an aspergilloma, or 'fungus ball'. In neutropenic patients, *Aspergillus* infection typically begins in the lungs and may be followed by fatal disseminated disease. Other sites, including the paranasal sinuses, skin, central nervous system and eye, may become infected; infection at these sites can have a poor prognosis.

Laboratory diagnosis

Sputum culture is of limited value. A positive isolate with bronchoalveolar lavage is diagnostic (98% specificity) but this approach lacks sensitivity.

The detection of precipitating antibody may confirm bronchopulmonary aspergillosis and farmer's lung, but is not helpful in immunocompromised patients as there is a poor antibody response. EIA to detect galactomannan in serial samples is useful. PCR has been described, but is not used routinely.

Treatment and prevention

Bronchopulmonary aspergillosis requires treatment of the airway obstruction with bronchodilators and steroids. Invasive aspergillosis requires systemic treatment with amphotericin B. Surgery may be beneficial in some cases of pulmonary infection. Patients with farmer's lung should avoid further exposure. Neutropenic patients should be nursed in rooms with air filters preventing inhalation of *Aspergillus* spores.

Other infections

Filamentous fungi may infect severely immunocompromised patients, the elderly, poorly controlled diabetics and chronic alcoholics. Sinus infection may spread to the eyes and brain. Pulmonary disease may be complicated by dissemination. *Mucor*, *Rhizopus* and *Absidia* are the main species implicated. Each is difficult to treat and has a poor prognosis.

Dermatophytes

Three species of filamentous fungi are implicated in dermatophytosis: *Epidermophyton*, *Microsporum* and *Trichophyton*. Dermatophytes are also grouped according to their reservoir and host preference: anthrophilic (mainly human pathogens); zoophilic (mainly infect animals); geophilic (found in soil and able to infect animals or humans). Anthrophilic species spread by close contact, e.g. families, enclosed communities. Transmission of geophilic species is rare. Close contact with animals may give rise to zoophilic infection (pet owners, farmers and vets).

Clinical features

Dermatophyte infection (ringworm) may present as red scaly patch-like lesions which spread outwards leaving a pale healed centre. Lesions are itchy but rarely painful although some species, notably zoophilic species, produce an intense inflammatory reaction with pustular lesions or an inflamed swelling (a kerion). Chronic nail infection produces discoloration and thickening whereas scalp infection is often associated with hair loss and scarring. Clinical diagnostic labels are based on the site of infection, e.g. tinea capitis (head and scalp), tinea corporis (trunk lesion).

Laboratory diagnosis

Infection of skin and hair by some species may demonstrate a characteristic fluorescence when examined under ultraviolet light (Wood's light).

Skin scrapings, nail clippings and hair samples should be sent dry to the laboratory. When heated in a solution of potassium hydroxide, they clarify and branching hyphae can be seen under the microscope. Dermatophytes grow on Sabouraud's agar at 30°C in 4 weeks.

Identification is based on colonial morphology, microscopic appearance, physiological and biochemical testing.

Treatment

Dermatophyte infections may be treated topically with imidazoles, e.g. clotrimazole. More troublesome infection may require oral treatment, e.g. terbinafine or griseofulvin.

Terbinafine

This inhibits squalene epoxidase with resultant accumulation of aberrant and toxic sterols in the cell wall. It is indicated for the oral treatment of superficial dermatophyte infections when infections are unlikely, or have failed to respond to local therapy. Stevens–Johnson syndrome and toxic epidermal necrolysis and hepatic toxicity are reported adverse effects. Treatment should be continued for up to 6 weeks for skin infections and 3 months or longer for nail infections.

Griseofulvin

It is active only against dermatophytes by inhibiting mitosis. Given orally, it is incorporated into the stratum corneum or nail where it inhibits fungal invasion of new skin and nail. Treatment must be continued until uninfected tissue grows. It is now rarely used.

Polyenes

There are two polyene cyclic macrolides in clinical use, nystatin and amphotericin B. They inhibit almost all fungi: amphotericin is also used to treat leishmaniasis (see Chapter 35). Polyenes are amphipathic molecules that bind to ergosterol in the fungal membrane forming a pore, which leads to leakage of the intracellular contents and cell death. Resistance is rare.

Nystatin is too toxic for systemic application: it is used for topical treatment and the prevention of fungal infection in immunocompromised patients. It has no value for the treatment of dermatophyte infections. Amphotericin can be given parenterally. The older formulations are relatively toxic, causing fever, chills, thrombophlebitis and hypotension. The drug also causes renal tubular damage that is usually reversible. Lipid formulations are much less toxic and higher doses can be safely given.

34 Intestinal protozoa

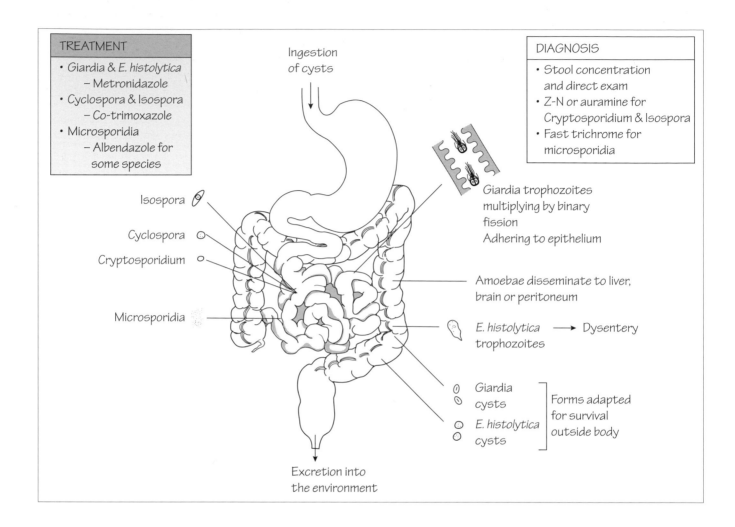

TREATMENT
- Giardia & E. histolytica
  – Metronidazole
- Cyclospora & Isospora
  – Co-trimoxazole
- Microsporidia
  – Albendazole for
    some species

DIAGNOSIS
- Stool concentration
  and direct exam
- Z-N or auramine for
  Cryptosporidium & Isospora
- Fast trichrome for
  microsporidia

Ingestion
of cysts

Isospora

Cyclospora

Cryptosporidium

Microsporidia

Giardia trophozoites
multiplying by binary
fission
Adhering to epithelium

Amoebae disseminate to liver,
brain or peritoneum

E. histolytica → Dysentery
trophozoites

Giardia
cysts
E. histolytica
cysts

Forms adapted
for survival
outside body

Excretion into
the environment

Entamoeba histolytica

Entamoeba histolytica infects the large intestine. Although more common in developing countries, cases are found world-wide. It is transmitted by the faeco–oral route and in food and water. Once ingested they adhere to the intestinal epithelium and produce proteases and amoebapore, an epithelial cytotoxin.

Clinical features

The onset is insidious with little systemic upset: the patient is ambulant but has frequent small-volume bloody stools with an offensive odour. Abscesses may develop in the liver or, more rarely, abdomen, lung or brain.

Diagnosis

Sigmoidoscopy reveals rectal ulceration: trophozoites can be demonstrated in ulcer biopsies. Three stool specimens should be sent where cysts can be identified; rarely, tropho-

zoites are found in fluid stools examined immediately. CT and ultrasound may reveal abscesses. Serology is useful in detecting abscesses but not intestinal infection.

Treatment

Metronidazole is effective in treating amoebic dysentery but does not eradicate the cyst stage, which requires dilox-anide furoate or paromomycin. Amoebic abscesses can usually be treated with metronidazole. Surgical drainage is usually unnecessary.

Prevention and control

Steps to ensure that water is boiled and food adequately cooked will reduce the risk of amoebic infection.

Giardia lamblia

Infection with *Giardia lamblia* is common throughout the world. It occurs where poor sanitation allows water supplies or food to be contaminated with *Giardia* cysts from human or possibly animal faeces.

Pathogenesis
Trophozoites multiply in the jejunum by binary fission. They attach themselves strongly to the intestinal wall by a sucking disk. The mechanism for *Giardia* diarrhoea remains unknown. The trophozoites pass down the intestine and develop into cysts, which are excreted in the faeces. This form is adapted for long-term survival.

Clinical features
Infection with *G. lamblia* is characterized by anorexia, crampy abdominal pain, borborygmus and flatus accompanied by bulky offensive fatty stools. Patients may lose weight and there may be an associated lactose intolerance or fat malabsorption. Infection in patients with IgA deficiency may suffer recurrent attacks of infection.

Laboratory diagnosis
Three stools should be examined and concentrated, as the shedding of *Giardia* cysts is intermittent. Aspirated jejunal contents can be examined immediately for the presence of motile trophozoites. Antigen detection tests are available.

Treatment
Metronidazole or tinidazole are used. Secondary malabsorption and vitamin deficiency may require investigation and treatment.

Cyclospora cayetanensis
This organism has only recently been recognized as a cause of human diarrhoea. Infection occurs world-wide and outbreaks related to contaminated water supplies and imported soft fruit have been reported in the USA.

Pathogenesis
Cyclospora are found inside vacuoles within the epithelium of the jejunum. There is inflammation, villous atrophy and crypt hyperplasia leading to malabsorption of B_{12}, folate, fat and D-xylose.

Clinical features
Infection takes the form of watery diarrhoea preceded by a flu-like illness and weight loss. It is self-limiting, but may last for weeks with continuing fatigue, anorexia and weight loss. In HIV-positive individuals disease is severe, prolonged and relapsing.

Diagnosis and treatment
Diagnosis is by demonstrating oocysts in stools. Co-trimoxazole is effective.

Cryptosporidium
Cryptosporidium parvum is a zoonotic coccidian parasite which is transmitted by milk, water and direct contact with farm animals. It is naturally resistant to chemical disinfectants, surviving water purification. Person-to-person spread can occur with intimate contact. Infection is common in children and HIV-positive individuals. It may interfere with the glucose-stimulated sodium pump in the small intestine, leading to fluid secretion.

Clinical features
Cryptosporidiosis is usually a self-limiting infection characterized by watery diarrhoea and abdominal cramps. In HIV-positive individuals diarrhoea is more profuse and may cause life-threatening fluid and electrolyte imbalance. They may also suffer relapse and infection of the biliary tree, gallbladder and respiratory tract.

Diagnosis and treatment
Cysts are demonstrated in the stool by Ziehl–Nielsen or auramine staining. There is no effective treatment; management is directed towards relief of symptoms.

Isospora belli
A coccidian parasite closely related to *Cryptosporidium*, *Isospora belli* presents with a similar clinical picture. There is often a history of travel to the tropics as it is not common in Europe. Ziehl–Nielsen's stain of stool identifies the characteristic oval cysts. Treatment is with co-trimoxazole.

Microsporidia
The microsporidia are small protozoan pathogens of insects, plants and animals. Organisms are intracellular, depending on host cells for a source of energy. They infect neighbouring cells using a long polar tube through which they inject their DNA. *Enterocytozoan beineusi*, *Encephalitozoan cuniculi*, *Encephalitozoan hellem*, *Septata intestinalis*, *Pleistophora* and *Nosema* have been implicated in human infection.

Pathogenesis
Enterocytozoan beineusi and *Septata intestinalis* infect epithelial cells of the small bowel, and are associated with diarrhoea. *Encephalitozoan cuniculi* infects macrophages, epithelial cells, vascular endothelial cells and renal tubular cells in the brain and the kidney. It is associated with hepatitis, peritonitis, diarrhoea, seizures and disseminated infection. Before the advent of HIV infection, microsporidia infection was very rare.

Diagnosis and treatment
Gram's stain, fast trichrome, calcofluor white and Ziehl–Nielsen have been used to demonstrate organisms. Albendazole may be effective in *S. intestinalis* infection.

35 Malaria, leishmaniasis and trypanosomiasis

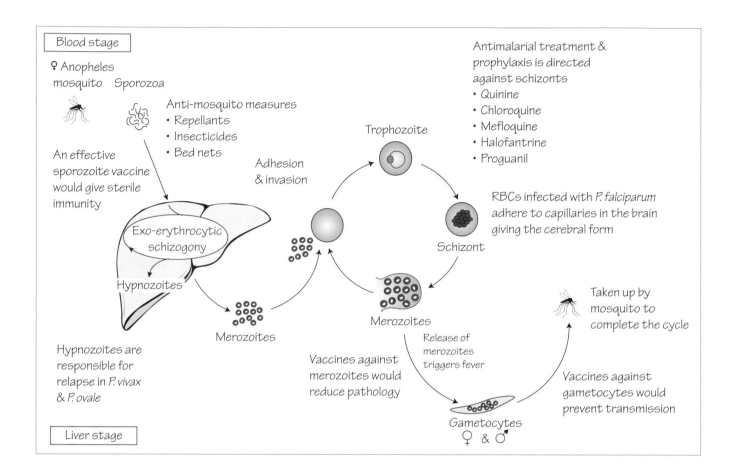

Malaria

Malaria is caused by four species of the genus *Plasmodium*: *P. falciparum*, *P. vivax*, *P. ovale* and *P. malariae*. More than 1.5 billion people live under the threat of malaria; 1 million children under the age of 5 years die each year in Africa alone. In the UK there are more than 2000 reported cases every year and up to 10 deaths.

Life-cycle

The infective stage, sporozoa, are injected into the human circulation by the bite of a female *Anopheles* mosquito. They invade hepatocytes, within which they replicate (exoerythrocytic schizogony). Some *P. vivax* or *P. ovale* parasites will develop into hypnozoites, which remain dormant; reactivation of hypnozoites will cause relapses. The parasites, now called merozoites, reinvade the circulation, enter the red blood cells (RBCs), multiply (erythrocytic schizogony) and, following rupture of the RBC, infect other RBCs. The parasites provoke the release of cytokines, which is responsible for many of the signs and symptoms of malaria. Infected RBCs develop knob-like projections and other antigens, making them adhere to the capillary wall. This may occur in the brain, causing cerebral malaria.

Some parasites differentiate into the sexual stages: the male and female gametocytes. These are taken up by female mosquitoes when they bite again and, after development in the mosquito gut, develop into the sporozoites which migrate to the insect salivary glands ready for another bite.

Clinical features

Malaria should be considered in any ill patient with a history of travel in an endemic area, particularly if they are feverish or have flu-like symptoms. Infection by *P. falciparum* can rapidly progress to death, especially in the non-immune traveller; infection by the other species is usually more benign. Holiday travellers have no immunity and regular fevers may not develop. *Plasmodium falciparum* affects every organ and so gives rise to a wide range of complications, e.g. cerebral malaria, circulatory shock, acute haemolysis and renal failure, hepatitis, and pulmonary oedema.

Diagnosis

At least three blood films (both thick and thin) should be obtained at different times, during or directly after a period of fever. Thick films improve the sensitivity of the test; thin films allow identification of the species and a parasite count to be performed.

Treatment

Drugs, such as chloroquine, kill the blood stages of the parasites. However, in recent years chloroquine-resistant *P. falciparum* has spread throughout the world and so *P. falciparum* infection is usually treated with quinine and Fansidar. Primaquine is also used for *P. vivax* and *P. ovale* infection to eradicate the hypnozoites.

Prevention and control

Sleep under bed nets. Cover exposed skin between dusk and dawn when mosquitoes are active. Use mosquito repellents. Take prophylaxis following expert up-to-date advice; remember that patients taking prophylaxis may still develop malaria.

Vaccines in development are directed mainly against the sporozoa, RBC stages and the gametocytes.

Leishmaniasis

Visceral disease is caused by *Leishmania donovani*, *L. infantum* or *L. chagasi*. Cutaneous disease is caused by several species, including *L. major*, *L. tropica* and *L. aethiopica* in the Old World and *L. braziliensis* and *L. mexicana* in the Americas.

Life-cycle

Leishmaniasis is transmitted by sandflies: *Phlebotomus* in the Old World and *Lutzomyi* in the Americas. Sandflies inject the infective promastigotes which survive ingestion by macrophages. They develop into the amastigote form, which multiply inside cells of the reticuloendothelial system.

Clinical features

Visceral disease is dominated by the effects of cytokines released by macrophages, e.g. tissue necrosis factor, giving rise to fever and general wasting. Bone marrow is replaced by parasites so the patient becomes anaemic, leucopenic and thrombocytopenic. There is an associated hypergammaglobulinaemia and patients are susceptible to secondary bacterial infections: untreated patients will deteriorate and die within 2 years.

Cutaneous forms are characterized by chronic granulomatous lesions at the site of the bite, although satellite lesions may also be present. *Leishmania braziliensis* infection causes cutaneous disease and, in some, destruction of structures around the mouth and nose.

Diagnosis and treatment

The demonstration of parasites in a skin biopsy, bone marrow sample, blood sample or splenic aspirate by microscopy and culture will confirm the clinical diagnosis. Both visceral and cutaneous leishmaniasis can be treated with liposomal amphotericin B.

Trypanosomiasis

African trypanosomiasis

African trypanosomiasis is caused by *Trypanosoma gambiense* and *Trypanosoma rhodesiense*. They are transmitted by the tsetse fly; humans are the only host of *T. gambiense* but antelope or cattle act as the reservoir for *T. rhodesiense*. Parasites enter the blood where the immune response initially reduces their numbers. However, the organism repeatedly changes its surface antigen and begins to multiply again. Generalized lymphadenopathy may be present and the skin may appear oedematous. The patient exhibits a hypergammaglobulinaemia and is susceptible to secondary bacterial infection. When parasites invade the brain, they cause a chronic progressive encephalitis: the patient lapses into coma and death is often the result of secondary bacterial pneumonia.

Diagnosis and treatment

Parasites may be demonstrated in blood samples or in lymph node aspirate. Serological tests are available. Parasites may be seen in samples of cerebrospinal fluid and associated with a characteristic 'morula cell'. Lumbar puncture should only be performed after circulating parasites have been eliminated with suramin, avoiding the risk of inoculation. The cerebral complications must be treated with Melarsoprol (MelB) which itself can cause serious toxicity.

South American trypanosomiasis

Trypanosoma cruzi is transmitted by the bite of reduviid bugs. There are three phases of the disease: acute infection characterized by cutaneous oedema, intermittent fever, shock and a significant mortality in children; latent infection; late manifestations, e.g. achalasia, megacolon, cardiac dysrhythmias, cardiomyopathy and neuropathy.

Diagnosis

The parasites may be demonstrated by microscopy, culture in artificial medium or may be introduced into suitable laboratory bugs (xenodiagnosis) which are later sacrificed. Serological tests are available.

Treatment

Nifurtimox may be used in the acute phase of infection. Treatment of complications is mainly palliative, e.g. cardiac pacemakers for heart block secondary to cardiomyopathy, surgery for megacolon.

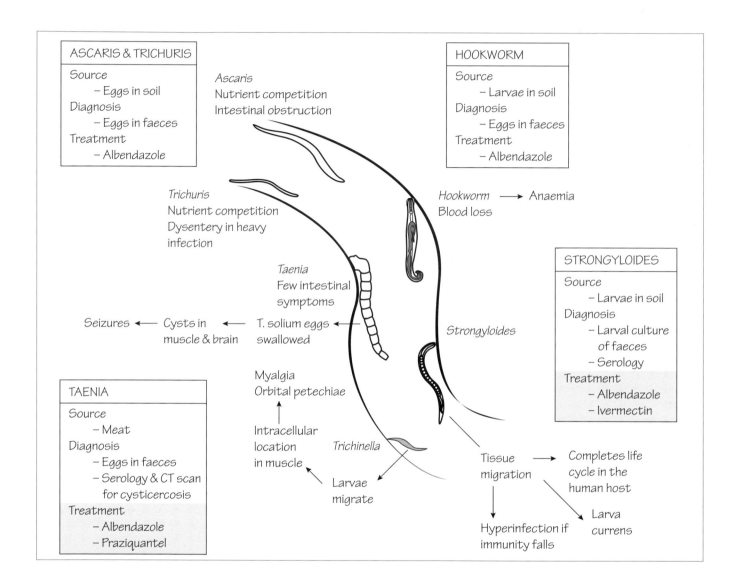

ASCARIS & TRICHURIS
Source
   – Eggs in soil
Diagnosis
   – Eggs in faeces
Treatment
   – Albendazole

Ascaris
Nutrient competition
Intestinal obstruction

HOOKWORM
Source
   – Larvae in soil
Diagnosis
   – Eggs in faeces
Treatment
   – Albendazole

Trichuris
Nutrient competition
Dysentery in heavy
infection

Hookworm → Anaemia
Blood loss

Taenia
Few intestinal
symptoms

Seizures ← Cysts in ← T. solium eggs ← Strongyloides
muscle & brain   swallowed

STRONGYLOIDES
Source
   – Larvae in soil
Diagnosis
   – Larval culture
      of faeces
   – Serology
Treatment
   – Albendazole
   – Ivermectin

Myalgia
Orbital petechiae

TAENIA
Source
   – Meat
Diagnosis
   – Eggs in faeces
   – Serology & CT scan
      for cysticercosis
Treatment
   – Albendazole
   – Praziquantel

Intracellular
location
in muscle

Trichinella

Larvae
migrate

Tissue → Completes life
migration   cycle in the
            human host

Hyperinfection if
immunity falls

Larva
currens

Roundworms and hookworms

Infection by nematodes (roundworms), *Ascaris lumbri-coides* and *Trichuris trichiura*, or the hookworms, *Necator americanus* and *Ancylostoma duodenale*, is very prevalent in developing countries. Both *Ascaris* and *Trichuris* are acquired by the ingestion of the roundworm eggs but hookworm larvae may invade the host through intact skin.

Epidemiology

Adult *Ascaris* are found in the intestine: the female worm produces up to 2000 eggs per day. The tough eggs survive in the soil where they mature into infective forms, which may be ingested. After hatching in the small intestine, the organism undergoes a migratory cycle through the liver and lungs, where it may be coughed up, eventually to develop into the adult worm resident in the intestine. Transmission occurs in conditions where there is poor sanitation or when food crops are manured with human faeces. In warm moist climates, the eggs can survive for many years in the soil. Hookworm eggs hatch an infective larva that is able to burrow through intact skin to cause infection.

Pathogenesis

These parasites cause disease by competing for nutrients; thus, the severity of symptoms is proportional to the number of worms present (parasite load). Hookworm infections are more serious as they take blood, leading to iron deficiency anaemia that can be severe. Heavily infected children have poor growth and lowered school performance, attributed to micronutrient deficiency.

Clinical features
Infection with intestinal nematodes is usually asymptomatic but a heavy *Ascaris* infection may lead to intestinal obstruction, and a heavy *Trichuris* infection to a dysentery-like syndrome.

Diagnosis
The diagnosis is made by examining up to three stool samples for the presence of the characteristic eggs.

Treatment
Intestinal nematodes can be treated with mebendazole or levamisole. Improved sanitation is required to control infection in the community.

Threadworms
Humans are the only host of *Enterobius vermicularis* (threadworms). They live in the large intestine, but the females migrate to the anus where they lay their eggs on the perianal skin. Infection is more common in children but often infects the whole family. Symptoms are few: thread-like worms may be found in the faeces or patients may complain of perianal itching, often worse at night. Scratching allows contamination of the fingers with larvae containing eggs which, when placed in the mouth, initiate a new cycle of infection. Occasionally, *Enterobius* can be found in the appendix.

Diagnosis is made by sending an adhesive tape swab to the laboratory where D-shaped eggs are seen. Treatment is with mebendazole or piperazine. It is often necessary to treat the whole family once, and again after 2 weeks.

Strongyloides stercoralis
Strongyloides stercoralis larvae are passed in the stools, where they either undergo a free-living cycle in the soil or differentiate into infective larvae that invade another host via intact skin. Inside the human host, they can initiate another development cycle. The consequence of this is that infection with *Strongyloides* can be prolonged. Resistance to *Strongyloides* depends on efficient cell-mediated immunity. Individuals who are infected with human T cell leukaemia virus I or are taking steroids are especially susceptible to hyperinfection.

Clinical features
Migrating larvae leave a red itchy track, which fades after about 48 h. If the patient is given immunosuppressive therapy, uncontrolled multiplication of the *Strongyloides* may occur, characterized by fever, shock and the signs of septicaemia and meningitis.

Diagnosis
Stool culture may reveal infective larvae. Alternatively, samples of jejunal fluid are examined for the presence of larvae. A sensitive EIA technique for serum is available.

Treatment
Ivermectin or albendazole are the main drugs used for treatment. Relapse occurs in up to 20% of patients. The hyperinfection syndrome is often accompanied by Gram-negative septicaemia, which requires vigorous treatment.

Prevention
The risk of infection can be reduced by wearing appropriate footwear to prevent larvae penetrating the skin.

Tapeworms
Two *Taenia* spp. infect humans: the pork worm, *Taenia solium*, and the beef worm, *Taenia saginata*. Infection is acquired by eating the meat of intermediate hosts that contains the tissue stages of the parasite.

Pathogenesis and clinical features
Tapeworms compete for nutrients and infections are usually asymptomatic.

Taenia solium can use humans as an intermediate as well as the definitive host. When an individual ingests *T. solium* eggs, they hatch and disseminate, forming multiple cyst-like lesions in the muscles, skin and brain. These 'measly' lesions, similar in appearance to infected pork meat, are known as cysticercosis. Inflammatory responses to parasitic antigens leaking from cysts in the brain may lead to epileptic seizures.

Diagnosis
This is made by finding characteristic eggs in the patient's stool. Cysticercosis is diagnosed by a specific EIA and confirmed by demonstrating the presence of multiple tissue cysts by X-ray, CT or MRI.

Treatment and prevention
Treatment is with praziquantel, but specialist advice should be sought for the management of *Taenia* infections in the central nervous system.

Diphyllobothrium latum
Humans are the definitive host of this tapeworm, acquiring infection from undercooked freshwater fish. The parasite competes for nutrients and causes deficiency of vitamin B_{12}. The diagnosis is made by detecting characteristic eggs in faeces. Treatment is with praziquantel.

Hymenolepis nana
Humans are the only host of this small tapeworm. Infection is usually asymptomatic and diagnosis is made by detecting the characteristic eggs in the faeces. Treatment is with praziquantel.

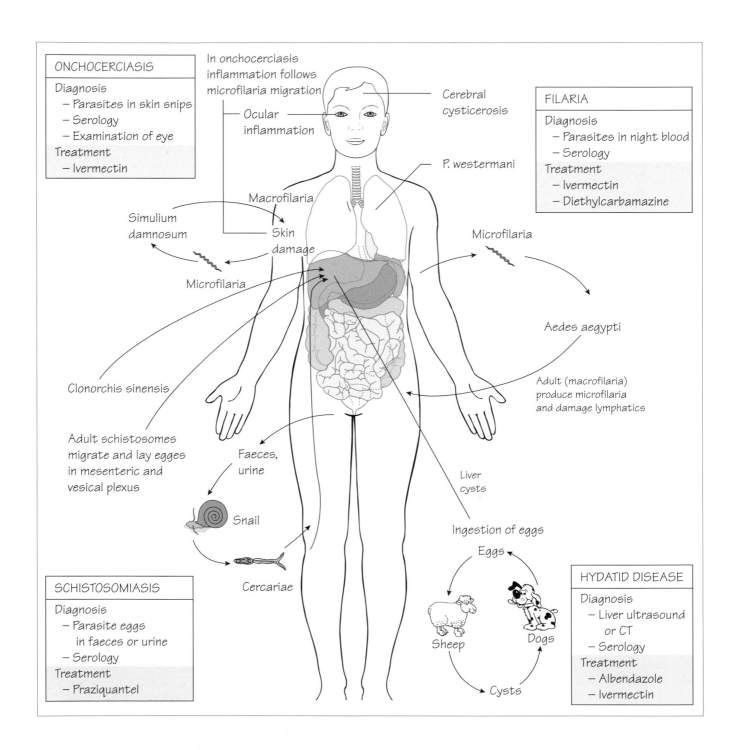

ONCHOCERCIASIS

Diagnosis
– Parasites in skin snips
– Serology
– Examination of eye
Treatment
– Ivermectin

In onchocerciasis inflammation follows microfilaria migration

Ocular inflammation

Cerebral cysticerosis

FILARIA

Diagnosis
– Parasites in night blood
– Serology
Treatment
– Ivermectin
– Diethylcarbamazine

Macrofilaria

P. westermani

Simulium damnosum

Skin damage

Microfilaria

Microfilaria

Aedes aegypti

Clonorchis sinensis

Adult (macrofilaria) produce microfilaria and damage lymphatics

Adult schistosomes migrate and lay eggs in mesenteric and vesical plexus

Faeces, urine

Liver cysts

Snail

Ingestion of eggs

Eggs

Cercariae

SCHISTOSOMIASIS

Diagnosis
– Parasite eggs in faeces or urine
– Serology
Treatment
– Praziquantel

Sheep

Dogs

Cysts

HYDATID DISEASE

Diagnosis
– Liver ultrasound or CT
– Serology
Treatment
– Albendazole
– Ivermectin

Schistosomiasis

Three species infect humans: *Schistosoma mansoni* (Africa and South America), *S. haematobium* (Africa) and *S. japonicum* (Far East). Eggs are excreted in the faeces and urine of infected humans. In areas of poor sanitation, the eggs hatch, releasing a miracidium; this invades the snail. After development, the schistosome cercariae emerge into the environment. They actively penetrate human skin, develop into male and female adult worms that migrate to the superior, inferior mesenteric or vesical plexus, depending on species.

Pathogenesis and clinical features

The first stage is characterized by fever, hepatosplenomegaly, skin rash and arthralgia. Associated with egg expulsion there may be bloody diarrhoea or haematuria. Later symptoms are caused by the fibrotic reaction to eggs, e.g. the liver (hepatic fibrosis and portal hypertension), the lungs and bladder (in the case of *S. haematobium*). Space-occupying lesions in the brain and spinal cord may lead to seizures.

Diagnosis

Eggs can be demonstrated in stool, urine, rectal snips or other tissue biopsy. An EIA is available.

Prevention and control

Infection is prevented by appropriate clothing when working in the fields and avoiding contaminated water. Control programmes targeting snails or mass treatment can control the disease if given sufficient resources.

Filaria

Classification

Lymphatic filaria: *Brugia malayi* and *Wuchereria bancrofti*; skin filaria: *Onchocerca volvulus*, and loiasis (*Loa loa*).

Epidemiology

Lymphatic filariasis is transmitted by the mosquito, *Aedes aegypti*, throughout the tropics. Onchocerciasis is transmitted by the black fly, *Simulium damnosum*, in West Africa and South and Central America. Loa is transmitted by *Chrysops* flies in West Africa.

Clinical features

Lymphatic filariasis is characterized by acute attacks of fever and lymphoedema, which may be complicated by secondary bacterial infection. After repeated attacks lymphatic vessels are permanently damaged leading to lymphoedema in the leg, arm or scrotum.

Onchocerca adults are located in nodules and microfilariae migrate in the skin resulting in pruritis and dry thickened skin. Inflammation in the eye causes blindness.

Loiasis is less damaging and diagnosis is based on fleeting subcutaneous swellings, known as Calabar swellings. Infection may be associated with fever and abnormalities of renal function.

Diagnosis

Lymphatic filariasis is diagnosed by identifying microfilariae in the peripheral blood at midnight. Blood is filtered and the filter stained and examined microscopically.

In onchocerciasis 'pinch biopsies' should be taken from any affected area, e.g. the shoulder blade, the buttocks and thighs. Examined microscopically, microfilariae can be seen to emerge from the skin. If negative, a 50-mg dose of diethylcarbamazine will induce increased itch. Pinch biopsies taken at this time are more likely to be positive. EIA is used for diagnosis.

Treatment

Lymphatic filariasis is treated with diethylcarbamazine and ivermectin is the treatment of choice for onchocerciasis.

Prevention and control

Onchocerciasis is an important public health problem in West Africa. An international control programme is underway, using mass treatment of whole populations with ivermectin. Lymphatic filariasis is prevented by mosquito control measures.

Hydatid disease

See Chapter 47

Clonorchis sinensis (Opisthorchis sinensis)

Infection is acquired by eating undercooked fish containing metacercariae. The adults live in the bile ducts and eggs are passed in the faeces. Infection is found mainly in the Far East, but can occur wherever infected fish is imported. Light infections are usually asymptomatic, but heavier infections result in cholangitis, pancreatitis, biliary obstruction and cirrhosis may develop. Cholangiocarcinoma is a late complication. The diagnosis is made by identifying the characteristic eggs in faeces. Patients may be treated with praziquantel. Infection is prevented by adequate cooking of potentially infected fish.

Fasciola hepatica

Humans are accidental hosts of this sheep and cattle parasite. The infective stage is found on freshwater plants, e.g. watercress, which, if eaten without cooking, can result in infection. The larvae hatch in the intestine: after maturation and migration, the adults are located in the liver. Patients present with fever and right upper quadrant pain, but these symptoms abate. Low-grade biliary symptoms and liver fibrosis may denote continuing infection. Treatment is with praziquantel.

Paragonimus spp.

Parasites of this genus infect different organs: *P. westermani*, the lungs; and *P. mexicanus*, the brain. Humans are infected by eating undercooked crustaceans. Acute non-specific symptoms, e.g. fever, abdominal pain and urticaria are followed by specific symptoms and signs, e.g. chest pain, dyspnoea and haemoptysis or central nervous system signs. Diagnosis is by identifying characteristic eggs in sputum, imaging, serology or tissue biopsy. Lung fluke is treated with praziquantel; cerebral disease with a combined surgical and medical approach.

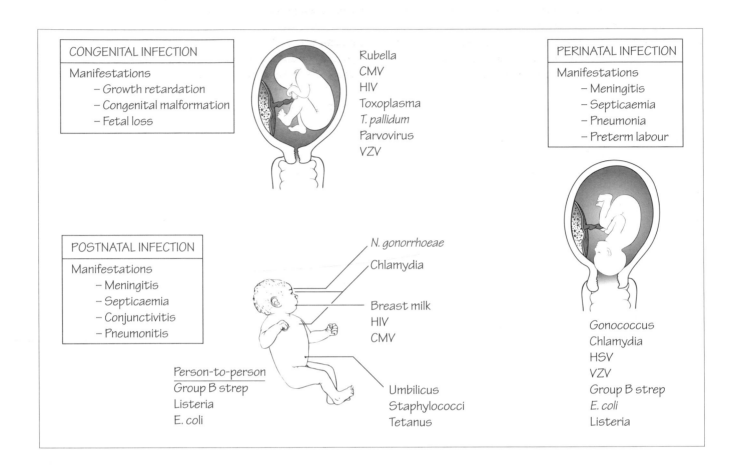

Maternal infection may cross the placenta causing intrauterine infection. Infection may be contracted during the process of birth, by direct contact with maternal blood or genital secretions. Prolonged rupture of the membranes predisposes to fetal infection. Infection can be transmitted to the neonate after birth.

Congenital rubella

Jaundice associated with hepatitis is often the first sign of congenital rubella. Haemolysis and thrombocytopenic purpura are also common, as is a low-grade meningo-encephalitis. Some babies have evidence of metaphyseal dysplasia. Infected infants have low birth weight and fail to attain their expected developmental milestones. There is a high mortality in severely affected infants. Patent ductus arteriosus, cataracts, deafness and retinal pigment dysplasia may be present.

Rubella IgM is positive and persists until the third month of life. As the risk of delivering an affected child is more than 60% if infected during the first trimester, some women will opt for termination of pregnancy. In later pregnancy, there is a balance between the likely fetal damage and the desirability of termination.

Cytomegalovirus

Infection is apparent at birth in about 10% of cases. It presents with prematurity, low birth weight, hepatomegaly, splenomegaly, thrombocytopenia and prolonged jaundice. About 25% of clinically affected neonates have cerebral irritability, fits or abnormal muscle tone or movement.

Microcephaly and sensorineural deafness are the most common problems. Other problems include cerebral calcification, hemiplegia, psychomotor retardation, choroidoretinitis and myopathy. Diagnosis depends on demonstrating IgM antibodies or cytomegalovirus excretion during the first 20 days of life.

Congenital and intrapartum herpes simplex infections

Primary herpes simplex infections may be accompanied by viraemia when transplacental infection can occur. Infants born with congenital infection tend to have severe disease, with pneumonitis, meningoencephalitis, hepatosplenomegaly and cytopenias. Only a few will demonstrate herpetic skin or mucosal lesions. Treatment with aciclovir reduces mortality from 80–90% to 10–15% and should not wait for laboratory confirmation.

Primary infection may be contracted at birth from maternal genital herpes. Skin, conjunctival, oral or genital lesions develop within a few days with dissemination in 50% of cases. Treatment is with intravenous aciclovir.

Varicella

Varicella embryopathy follows maternal infection during the first or second trimester of pregnancy; it is transmitted in less than 3% of infected pregnancies. Cicatricial contracture of a limb with hypoplasia, microcephaly or microphthalmia may occur. Non-immune women exposed to chickenpox should be offered post-exposure prophylaxis with zoster immune globulin (ZIG) within 10 days of exposure.

Neonatal varicella occurs when the mother develops chickenpox within 1 week of delivery. As neonatal mortality is up to 40%, the neonate should be given ZIG as soon after birth as possible. If the mother's disease appears within 1 week after birth, ZIG should also be given to the infant as soon as possible, preferably within 48 h. Immunoglobulin given to the mother will not protect the infant. A vaccine is entering clinical use in some countries.

Gonococcal ophthalmia neonatorum

Neisseria gonorrhoeae infection may be contracted during delivery causing ophthalmia neonatorum, a purulent conjunctivitis. It is diagnosed by direct Gram's stain and culture. Chloramphenicol ointment will treat the infection.

Listeriosis

Transplacental transmission of *Listeria monocytogenes* occurs during a maternal infection that is often inapparent. Infection in early pregnancy often results in fetal death; later infection is associated with premature labour. Severe bacteraemia, associated with hepatosplenomegaly, meningoencephalitis, thrombocytopenia and pneumonitis, usually complicates neonatal infection. Intrapartum exposure may lead to neonatal infection during the first 2 weeks of life, usually with meningitis and bacteraemia. Blood and CSF, placental tissue and lochia should be cultured. Infected mothers and infants may be a source of infections in the postnatal ward, and should be isolated. Ampicillin and aminoglycosides are the treatment of choice: treatment should continue for up to 3 weeks to prevent relapse.

Syphilis

Congenital infection is now rare as a result of antenatal screening. Affected babies are feverish with features similar to secondary syphilis: rash, condylomata and mucosal fis-

sures. Osteochondritis may cause pain. Persistent rhinitis ('snuffles') is common.

Diagnosis is confirmed by dark-ground microscopy of mucosal or skin lesions. Specific IgM or antibodies persisting after 6 months indicate infection. Late manifestations appear between 12 and 20 years: deafness, optic atrophy or paretic neurosyphilis. Other features include bossing of the frontal bones, chronic tibial periostitis, notching of the incisors, 'mulberry' deformity of the first permanent molar and a high arched palate. The treatment of choice is benzylpenicillin.

Toxoplasmosis

The incidence of toxoplasmosis varies internationally: it is uncommon in the UK, but common in France. Transplacental infection occurs in a third of affected pregnancies. Infection in the first and second trimester is more likely to cause significant fetal disease: the fetus may be stillborn, die soon after birth, or have cerebral calcification, cerebral palsy or epilepsy. Chorioretinitis may not be evident until after birth and may be the only feature. Maternal toxoplasmosis is confirmed by specific IgM antibodies or by seroconversion. IgM antibodies may also be demonstrated in affected neonates. Treatment with spiramycin may reduce the risk of transplacental infection but does not effect the outcome of fetal disease.

Chlamydia

Chlamydial ophthalmia neonatorum is a severe conjunctivitis appearing within 4 days of birth. It is often followed, at 6 weeks of age, by pneumonitis characterized by tachypnoea and cough. Conjunctivitis is treated with topical tetracycline. Erythromycin is the treatment of choice for chlamydial pneumonitis.

Perinatal infections
Bacteraemia and pneumonia
In the first few days of life there are few specific clinical features of bacteraemia. The neutrophil count may rise, though this is not always reliable. Meningitis also presents nonspecifically. Blood, urine and CSF culture should be performed but treatment should not wait for laboratory confirmation.

Therapy should be targeted at *E. coli* and group B streptococci, e.g. benzylpenicillin and gentamicin, or cefotaxime.

Bullous impetigo (Lyell's syndrome)
Infection with *Staphylococcus aureus* expressing exfoliative toxins results in superficial blisters or bullae which break leaving extensive raw areas—the 'scalded skin syndrome'.

Treatment with flucloxacillin should be prescribed. The infant and mother should be isolated.

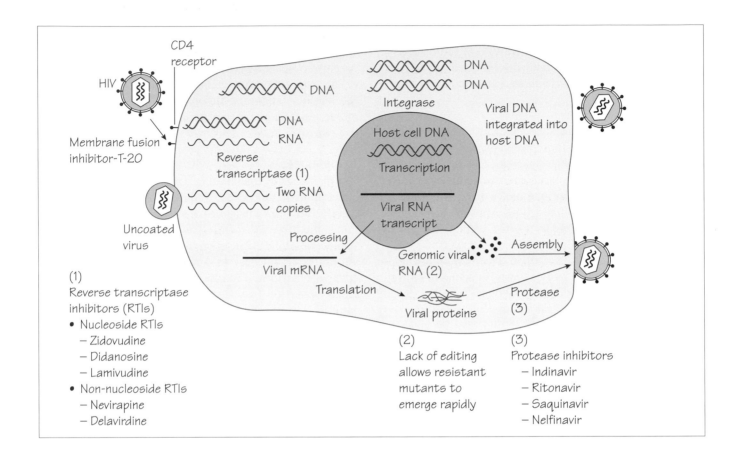

HIV

HIV is a spherical enveloped RNA virus. It is a retrovirus, so-called because it possesses reverse transcriptase, which produces a DNA copy from viral RNA that is incorporated into the host nucleus to become the template for further viral RNA. Three genes are required for viral replication—*gag*, *pol* and *env*. HIV is classified as a lentivirus and two are pathogenic for humans: HIV-1, which is most common; and HIV-2 which is found mainly in West Africa and appears to be less virulent.

Epidemiology

HIV infection has spread world-wide, transmitted by the parenteral and sexual routes. Infection is most common in patients at high risk of sexually transmitted disease, especially where genital ulceration is common. In developed countries, the main risk groups are homosexuals and intravenous drug abusers. In developing countries HIV spreads mainly by heterosexual transmission and through unscreened transfusions or contaminated medical equipment. Infection can be transmitted from mother to fetus.

Pathogenesis

The virus infects cells with a CD4 receptor, e.g. T cells, and macrophages. Viral replication results in the progressive T cell depletion and diminished cell-mediated immunity. Lacking T cell help, B cell function is also reduced. HIV causes damage to neural cells and stimulates cytokine release that may also cause neurological damage. Many of the clinical signs of HIV infection are caused by secondary infections.

Clinical features

A few weeks after infection, a mononucleosis-like syndrome may develop with rash, fever and lymphadenopathy. A latent period follows that may last as long as 10–15 years. When T cell function is sufficiently compromised, secondary infections and malignancies develop as a result of profound immunosuppression (although this is reversed by highly active antiretroviral therapy (HAART)):

Bacterial: *M. tuberculosis*, *M. avium-intracellulare*, *Salmonella*, *Streptococcus pneumoniae*.

Protozoan: *Toxoplasma gondii*, *Cryptosporidium parvum*, *Isospora belli*, Microspora.

Fungal: *Candida* spp., *Cryptococcus neoformans*, *Pneumocystis carinii*.

Viruses: varicella zoster, JC human papovavirus.

Malignancy: Kaposi's sarcoma (now considered a human herpesvirus infection), non-Hodgkin's lymphoma.

Children with HIV are especially vulnerable to childhood virus infections, e.g. measles, and recurrent bacterial infections, e.g. pneumonia. A common presentation is failure to thrive.

Diagnosis

Diagnosis is by detection of HIV-specific antibody using two different immunoassay methods, e.g. EIA, competitive EIA, particle agglutination or Western blotting. Individual testing must always be preceded by counselling. In addition, as seroconversion may take up to 3 months, an initial negative result should be repeated.

PCR can be used to detect HIV and RNA in clinical samples and quantitative assays measure the viral load, which can be used to monitor treatment. HIV may be grown in lymphocyte cultures but this not used in diagnosis.

Treatment

Agents available to treat HIV infection include nucleoside reverse transcriptase inhibitors (NTRIs), e.g. zidovudine; non-nucleoside reverse transcriptase inhibitors (NNRTIs), e.g. nevirapine; and protease inhibitors, e.g. indinavir. At present, three main principles govern treatment.

1 Initial treatment, with at least three drugs, should start before substantial immunodeficiency develops.

2 When therapy changes, at least two drugs should be added or substituted to prevent resistance emerging.

3 As treatment progresses, the viral load should be monitored to keep it below the level of detection by current assays.

Because RNA viruses lack efficient proof-reading, mutations arise rapidly and patients develop drug resistance quickly. Resistance and drug intolerance are the main reasons for stopping therapy.

Prevention

HIV transmission is prevented by avoiding high-risk partners and unprotected intercourse, e.g. using barrier contraception. Blood products must be screened and potentially HIV-infected material discarded. Health education and free needle exchange programmes may reduce the risk of transmission between drug abusers. Vaccine development is hindered by the antigenic variability shown by the virus.

Pneumocystis carinii

Pneumocystis carinii, once considered a protozoan, is probably more closely related to fungi. Infection only occurs in patients who have severe T cell dysfunction through HIV, malnutrition, prematurity, primary immune deficiency diseases and immunosuppressive drugs. Prior to the HIV epidemic, the infection was rare. Transmitted by the respiratory route, *P. carinii* adheres strongly to pneumocytes.

Clinical presentation

Patients typically present with dyspnoea, which develops insidiously over days or weeks, and an unproductive dry cough. Pleuritic chest pain is uncommon. Although patients are febrile, clinical examination is usually normal, although fine basal crackles may be heard.

Initially the chest X-ray may appear to be normal but reticular shadowing may develop until there is diffuse air space consolidation. In a small proportion of patients there are atypical features.

Diagnosis

Specimens, obtained by bronchoalveolar lavage or by the use of nebulized hypertonic saline, are examined by specific immunofluorescence, methenamine silver staining or PCR.

Treatment

Treatment is with oral co-trimoxazole in high dosage, or intravenous pentamidine. Alternatives include trimethoprim–dapsone, pyrimethamine–clindamycin and atovaquone.

Fungal infections

HIV patients may develop severe mucocutaneous candidiasis with oral ulceration and oesophageal infection. This causes dysphagia with subsequent weight loss. Acute infection is treated with oral agents, e.g. fluconazole, but as long-term suppressive treatment is required, resistance to these agents may develop. Cryptococcal meningitis is a common and recurrent problem (see Chapter 42).

Mycobacterial infections

HIV-positive individuals are at increased risk of tuberculosis and *Mycobacterium avium-intracellulare* (see Chapter 13).

Toxoplasma gondii

Toxoplasma infection persists inside the host cells for very long periods. Falling immunity (CD4 count $<0.05 \times 10^9$/litre) allows the reactivation of previous dormant infection. A space-occupying lesion may develop in the brain that may be accompanied by encephalitis.

Toxoplasma encephalitis presents with fever and headaches. Convulsions, coma and focal neurological signs may follow. Computed tomography scan may demonstrate multiple diagnostic focal lesions with ring enhancement. Brain biopsy may yield material for tissue culture or PCR. *Toxoplasma* encephalitis is treated with pyrimethamine–sulfadiazine. Long-term suppressive treatment is required after recovery.

Pyrexia of unknown origin and septicaemia

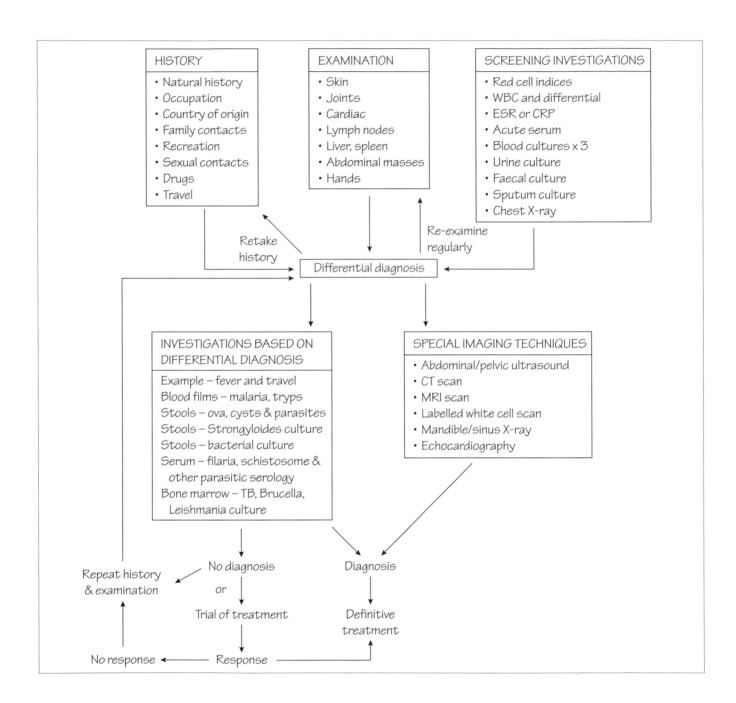

Pyrexia of uncertain origin

Definition

A fever, intermittent or persistent, of greater than 38.2°C for more than 2 weeks for which there is no obvious cause, is termed pyrexia of unknown origin (PUO).

Aetiology

Infection accounts for 45–55% of cases of PUO. Malig-

nancy accounts for 12–20% of cases: common causes are lymphomas, renal cell or lung carcinomas. Connective tissue disorders, such as rheumatoid arthritis, systemic lupus erythematosus or polyarteritis nodosum, are responsible for 10–15% of cases. Hypersensitivity to drugs, pulmonary emboli, granulomatous diseases (e.g. sarcoidosis), rare metabolic conditions (e.g. porphyria) and factitious (i.e. fever induced deliberately by the patient) are less common causes that should be considered. The longer the history of fever, the more likely it is that it arises from a *non*-infectious cause.

Investigation

When investigating a possible infective cause, a detailed history of the presenting complaint including the occupational, social and sexual history of the patient is important, e.g. wind-surfers are at risk of leptospirosis; veterinarians and farmers of zoonotic infections, e.g. brucellosis. Recent travel may suggest exposure to unusual tropical infections, some of which have a long incubation period. Sexual history may indicate possible HIV or related risks. The medical history should include a detailed list of medications, whether prescribed by a doctor or not. Many drugs and health products can cause fever. Patients may be reluctant to tell their doctor about drugs purchased from an alternative medicine practitioner. Patients may overlook vital information at the first interview: further opportunities for discussion are often necessary.

The patient must be examined carefully for localized bone or joint pain, subtle rashes, lymphadenopathy, abdominal masses, cardiac murmurs and mild meningism. A complete physical examination should be repeated regularly to detect changes, such as a new soft systolic murmur or increasing abdominal tenderness. A spleen or liver which was impalpable on first examination may have become so because of the development of the condition.

After the initial history and examination is taken the investigative process can be divided into three phases.

1 Screening investigations which are performed on all patients (see diagram).

2 Investigations which depend on the results of history and physical examination.

3 Further screening tests and special imaging techniques: ultrasound, computed tomography, magnetic resonance imaging, echocardiography and dental X-ray.

The results of the preliminary history and examination are taken together with the results of the primary investigations to plan the tests that will be performed in the second round. These investigations are chosen based on syndrome groups, e.g. fever, eosinophilia and tropical travel. If a diagnosis is not made, further imaging techniques may reveal an occult abdominal abscess or osteomyelitis.

Management

A diagnosis should be made before any antimicrobial therapy is commenced. In some infections, notably tuberculosis, which are suspected but not proved, a trial of therapy may be considered. If this produces clinical improvement, a full course of chemotherapy may be initiated.

Septicaemia

Aetiology

Bacteraemia may arise from normal flora, which have become invasive in conditions such as dental abscess, cholecystitis, appendicitis or diverticulitis. Septicaemia following surgery may be caused by a wide range of organisms, including contamination by skin flora. This problem is particularly important in surgery involving prosthetic devices (orthopaedic, cardiovascular, neurosurgical). The urinary tract is a very common source of Gram-negative infection (see Chapter 44). *Streptococcus pneumoniae* bacteraemia (see Chapter 43) may follow pneumonia; *Streptococcus pyogenes* or *Staphylococcus aureus* bacteraemia may complicate skin infections. Septicaemia caused by *Neisseria meningitidis* or *Streptococcus pneumoniae* may be accompanied by meningitis.

Clinical features

Although the septicaemic patient is usually severely ill with fever and shock, sometimes aggravated by depressed consciousness, septicaemia may be asymptomatic. Fever may be absent in children and the elderly, shock may not have yet developed, and they may present as confused, drowsy or generally unwell. Clinically, it is impossible to distinguish Gram-positive from Gram-negative shock. Some organisms may have characteristic associated clinical signs, e.g. the purpura of *N. meningitidis*.

Diagnosis and treatment

With the exception of suspected *N. meningitidis* infection, at least two specimens of peripheral blood should be taken for culture before therapy is commenced. Other investigations are directed to finding the source of the sepsis, e.g. urine culture, cerebrospinal fluid, abdominal ultrasound, sputum culture and skin swabs. Chest and abdominal X-ray should be performed.

Empirical therapy based on the source of infection and likely infecting organism should be started promptly. Parenteral antibiotics with a broad spectrum, covering the likely pathogens, should be prescribed.

Puerperal fever

This is a severe, usually bacteraemic, infection caused by entry of pathogens through the placental bed or the cervix within 7 days of delivery. Fever, back pain, offensive lochia or shock may be present. Infection may be complicated by disseminated intravascular coagulation. Fever in the early puerperium should be investigated with blood and urine culture and endocervical swabs. Empirical treatment with a third-generation cephalosporin and metronidazole should begin without delay. Retained products of conception should be removed. Intensive care support may be required.

41 Endocarditis, myocarditis and pericarditis

INFECTIVE ENDOCARDITIS (IE) – AETIOLOGY

Native valve
• Viridans group streptococci
• Enterococci
• Other streptococci
• Staphylococcus aureus
• Coagulase-negative staphylococci
• Fastidious Gram-negatives

Culture negative endocarditis
• NB serological diagnosis
• Previous antibiotic therapy
• Chlamydia pneumoniae/psittaci
• Coxiella burnetti (Q fever)
• Mycoplasma

Prosthetic valve – Early
• Coagulase-negative staphylococci
• Staphylococcus aureus
• Viridans group streptococci
• Enterococci and other streptococci
• Fungi
Late – as for native valve

Right sided
• Nutritionally deficient strains
• Staphylococcus aureus
• Mixed infections
• Fungi

IE – PREDISPOSING FACTORS

• Atherosclerosis/ischaemic changes
• Degenerative changes
• Congenital abnormalities, e.g. VSD, coarctation
• Rheumatic fever
• Prosthetic material, e.g. valves, pacing wires, patches/grafts, central venous lines
• IV drug abusers – right sided

PERICARDITIS – AETIOLOGY

• Pneumonia, e.g.– Pneumococcus
   Staphylococcus aureus
   M. tuberculosis
• Enterovirus
• Influenza
• Mycoplasma

IE – PATHOGENESIS

Damage and roughening of endothelium
↓
Fibrin and platelet deposition
   Bacteraemia – oropharynx/gut/urinary tract
Colonization of deposit
↓
Bacterial multiplication, further fibrin and platelet deposition, immune activation
↓
Systemic signs of infection, development of vegetation, toxic, embolic and immune complex phenomena

MYOCARDITIS – AETIOLOGY

• Coxsackievirus
• Echovirus
• Adenovirus
• Rubella
• Mycoplasma
• Toxic – diphtheria
      – septicaemia
NB Immune mediated

Endocarditis

Heart valves may be infected during transient bacteraemia. Congenitally abnormal or damaged valves are at greatest risk. Bacteria may originate from the mouth, urinary tract, intravenous drug misuse or colonized intravascular lines.

Clinical features

Patients present with malaise, fever and variable heart murmurs. Arthralgia is sometimes present. The classical stigmata, e.g. splinter haemorrhages, Osler's nodes, microhaematuria, retinal infarcts, finger clubbing, *café-au-lait* skin, Janeway's lesions, are only seen when infection has been present for some time. In later stages, septic emboli may cause a stroke. With more virulent organisms, *Staphy-*

lococcus aureus, infection progresses rapidly and signs of acute sepsis may predominate.

Complications
Local progression may lead to abscess formation in the aortic root. Destruction of the valve results in rapid cardiac decompensation and severe cardiac failure. Cerebral or limb infarction may follow septic embolus. Nephritis is secondary to immune complex deposition and can progress rapidly if sepsis is uncontrolled or renal-toxic antibiotics are given, e.g. aminoglycosides.

Investigation
Echocardiography will demonstrate vegetations on the valves; a plain chest X-ray may show evidence of cardiac failure. At least three sets of blood cultures should be taken, an hour apart, while fever is present. Antibiotic therapy should await the results of blood culture if possible. Serum should be tested for antibodies to *Coxiella* and *Chlamydia psittaci*.

Management
Ideally, antibiotics should not be commenced until the identity and sensitivities of the infecting organism are known; the prognosis of empirically treated, culture-negative endocarditis is poorer than when the infecting organism is identified. Careful microbiological monitoring of the markers of inflammation, e.g. CRP, is associated with an improved outcome.

Therapy should be planned based on sensitivity testing and following determination of minimum inhibitory concentration (MIC) and the minimum bactericidal concentration (MBC). Gentamicin levels must be closely monitored because patients with endocarditis are particularly susceptible to the toxic effects as a result of renal impairment. Therapy is continued for 2–6 weeks. Typical regimens include benzylpenicillin and gentamicin for streptococci viridans; flucloxacillin and gentamicin for staphylococci; vancomycin; gentamicin for penicillin-allergic patients.

Surgical management may be required to deal with the haemodynamic consequences of endocarditis, especially in cases caused by *S. aureus* and other more virulent pathogens, or if infection is unresponsive to antimicrobial therapy.

Prevention
Endocarditis may be prevented by giving antibiotic prophylaxis to patients with damaged valves when they undergo procedures which give rise to significant bacter-aemia, e.g. dental work or urogenital surgery (3 g amoxicillin). If the urine is infected, the antibiotic choice should reflect the sensitivity of the urinary organism cultured. For procedures requiring an anaesthetic, prophylaxis is given at induction (1 g amoxicillin i.m., followed by 0.5 g orally 6 h later). There are alternative regimens for penicillin allergy and prosthetic valves.

Myocarditis
Most solitary myocarditis is caused by viral infection. However, it may complicate systemic viral infections, follow bacteraemia or form part of brucellosis, rickettsial or chronic Chagas' infection.

Patients present with influenza-like symptoms associated with fatigue, exertional dyspnoea, palpitations and precordial pain. Tachycardia, dysrhythmia or cardiac failure may be present. The ECG may show T-wave inversion, prolongation of the PR or QRS interval, extrasystoles or heart block. There may be an elevation in cardiac enzymes and cardiomegaly on chest X-ray.

The diagnosis is suggested by the relationship of viral symptoms to the development of cardiological abnormalities. Enteroviruses may be recovered from throat or stool cultures, or respiratory viruses from nasopharyngeal or throat specimens (see Chapter 29). Treatment is supportive.

Pericarditis
Pericarditis is most often secondary to a non-infectious condition, e.g. myocardial infarction. It may also arise as a complication of bacteraemia, following spread of pus from an empyema (*S. pneumoniae*) or from a liver abscess (enterococci, *Entamoeba histolytica*). Tuberculosis can cause subacute pericarditis.

Viral pericarditis is a self-limiting condition featuring fever, flu-like symptoms and sharp chest pain. Enteroviruses, especially coxsackie and influenza viruses, are most commonly implicated. The chest pain may vary with posture, swallowing or heartbeat. A pericardial rub may be heard. Cardiographic evidence of pericarditis may be demonstrated.

Patients with suppurative pericarditis present with fever, neutrophilia and signs of the underlying source of infection. Chest pain is severe and a fall in blood pressure may indicate developing tamponade. Electrocardiographic changes show upward-curved elevated ST segments. Echocardiography will show pericardial thickening or effusion. Infection can be complicated by fibrosis and constrictive pericarditis, leading to congestive cardiac failure. Treatment is directed towards the likely causative organism.

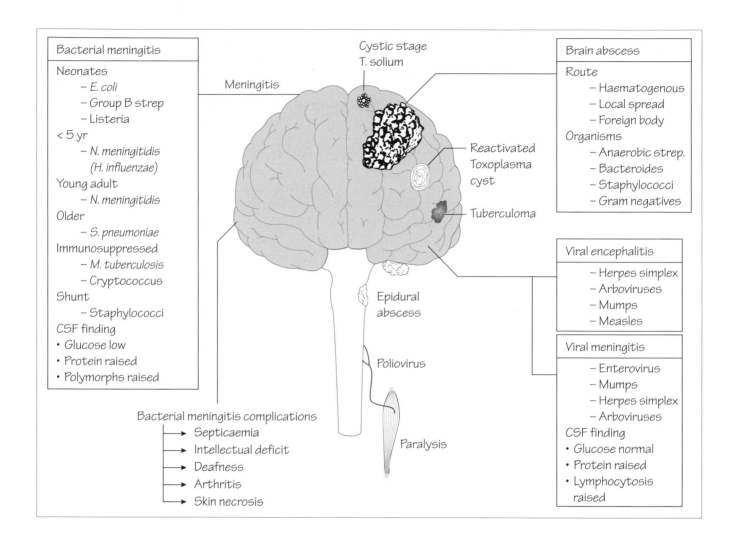

Meningitis

Aetiology

A wide range of organisms may cause bacterial meningitis: different organisms are more frequent at different ages and patient groups (see diagram).

Clinical features

Classically, meningitis presents with fever, headache, photophobia and neck stiffness. Vomiting and diarrhoea may predominate. The level of consciousness progressively falls. Not all these signs and symptoms may be present, especially in neonates and the elderly who often present atypically. A bulging fontanelle in a neonate indicates raised intracranial pressure. Ear or sinus infections may indicate pneumococcal disease.

Haemophilus influenzae meningitis can be complicated by recurrence of fever, hydrocephalus, convulsions and deafness. *Neisseria meningitidis* can be complicated by deafness, intellectual deficit, skin necrosis and reactive arthritis. With appropriate treatment, mortality from *H. influenzae* and *N. meningitidis* should be less than 5%. However, the onset of the purpuric rash from *N. meningitidis* septicaemia can be rapidly fatal. Sequelae including deafness, cranial nerve palsies and hydrocephalus are most frequent following *S. pneumoniae* meningitis, which also has the highest mortality (20%).

Diagnosis

Unless there is clinical suspicion of raised intracranial pressure, a sample of CSF should be obtained. For total and differential white cell count, Gram stains, Ziehl-Nielsen, india ink, PCR and rapid antigen detection. Bacterial meningitis causes a high white cell count, predominantly neutrophils, a low glucose and raised protein. In tuberculous meningitis there is a lymphocyte response with high protein and low glucose. Blood should be taken for culture, PCR and rapid antigen detection, and comparative glucose.

Management
Neonatal meningitis may be treated with cefotaxime and an aminoglycoside. Ampicillin must be added if *Listeria* is suspected. *Haemophilus* meningitis requires cefotaxime. *Neisseria meningitidis* is invariably susceptible to penicillin. Penicillin-resistant *S. pneumoniae* is now being reported and cefotaxime can be substituted. Cryptococcal meningitis is treated with amphotericin and 5-flucytosine. Tuberculous meningitis is treated with rifampicin, pyrazinamide isomazid and ethambutol (see Chapter 13).

Prevention
Capsular polysaccharide vaccines are available for *N. meningitidis* serogroups A, C and W135 but not serogroup B which causes most cases in the UK. Conjugate vaccines against A and C have recently become available. Similar vaccines against *H. influenzae* meningitis, have been very successful. Family (close) contacts of meningococcal and *Haemophilus* meningitis patients require antimicrobial prophylaxis (ciprofloxacin or rifampicin).

Brain abscess
Brain abscesses arise from parameningeal suppuration, foreign bodies or haematogenous spread from distant sepsis. Infection is often mixed anaerobic cocci, e.g. *Prevotella* spp., *Bacteroides* spp., *Fusobacterium* spp., staphylococci, streptococci (*S. milleri*) and enterobacteria.

Clinical features
Patients present with headache, fever, and lowered consciousness. Focal neurological signs depend on the location of the abscess. Signs of raised intracranial pressure may develop (rising blood pressure, falling pulse) followed by seizures.

Diagnosis and treatment
Lesions are localized by CT scanning. A lumbar puncture may be dangerous because of the risk of cerebellar herniation. Drainage should be performed if feasible and pus sent for culture and sensitivity testing.

In addition, a regimen of cefotaxime and metronidazole or a combination of benzylpenicillin, chloramphenicol and metronidazole may be used.

Viral meningitis
Meningitis and encephalitis may arise from infection with enteroviruses, mumps, herpes simplex, arboviruses, influenza and, rarely, rubella or Epstein–Barr virus. Viral meningitis can be part of the natural history of polio infection (see Chapter 29). Patients present with headache, photophobia, fever and neck stiffness. The CSF shows an increase in lymphocytes, the protein is mildly raised with normal glucose levels. Throat swabs, CSF and stool specimens should be sent for viral culture and serological testing. Management is symptomatic as most patients recover without residual deficit within a few days.

Viral encephalitis
Viral encephalitis is caused by a variety of viruses. Patients are febrile with headache, neck stiffness, and impaired consciousness. Focal neurological signs may develop: convulsions are common. Virus may be cultured from CSF, stool and throat specimens, and detected by serological techniques. Aciclovir is used for treatment of herpetic encephalitis (which typically affects the temporal lobe) reducing both the mortality rate to less than 20%, and the number of patients with severe residual disability.

Postinfectious encephalitis
A number of viruses are associated with encephalitis that arises after the systemic infection has resolved (postinfectious encephalitis): measles, varicella-zoster, rubella, Epstein–Barr virus, mumps and influenza. Clinically similar to viral meningitis, it is thought to be mediated by an autoimmune reaction.

Spongiform encephalopathies
The prion protein is a protease-resistant form of a protein that is a normal constituent of the brain. When ingested, the prion protein induces a conformational change in the host brain protein, leading to spongiform degeneration in the brain. There is an extended incubation period of more than 5 years.

Kuru was described in cannibals from Papua New Guinea who ate human tissue, including brain. Recently, bovine spongiform encephalopathy, (BSE) is thought to have been caused by feeding animal brain protein to cattle. Transmission to humans, following ingestion of contaminated beef products, is thought to be responsible for variant Creutzfeld–Jakob disease. The incidence of BSE has diminished rapidly because of a selective slaughter policy and a ban on feeding animal protein to cattle. The size and scale of any human epidemic is not yet known.

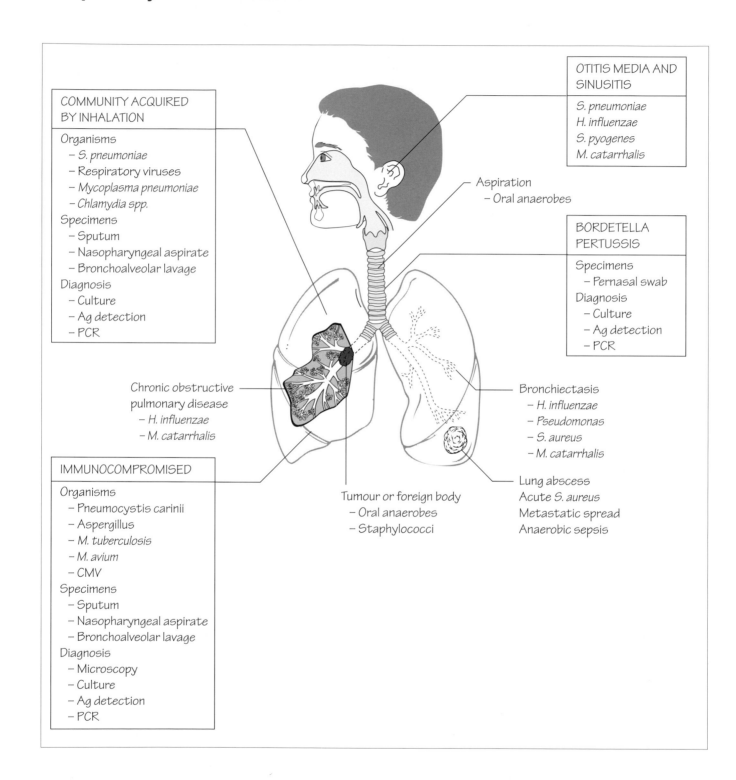

COMMUNITY ACQUIRED
BY INHALATION

Organisms
  – *S. pneumoniae*
  – Respiratory viruses
  – *Mycoplasma pneumoniae*
  – *Chlamydia spp.*
Specimens
  – Sputum
  – Nasopharyngeal aspirate
  – Bronchoalveolar lavage
Diagnosis
  – Culture
  – Ag detection
  – PCR

OTITIS MEDIA AND
SINUSITIS
  S. pneumoniae
  H. influenzae
  S. pyogenes
  M. catarrhalis

Aspiration
  – Oral anaerobes

BORDETELLA
PERTUSSIS
  Specimens
    – Pernasal swab
  Diagnosis
    – Culture
    – Ag detection
    – PCR

Chronic obstructive
pulmonary disease
  – *H. influenzae*
  – *M. catarrhalis*

Bronchiectasis
  – *H. influenzae*
  – *Pseudomonas*
  – *S. aureus*
  – *M. catarrhalis*

Lung abscess
Acute *S. aureus*
Metastatic spread
Anaerobic sepsis

IMMUNOCOMPROMISED

Organisms
  – Pneumocystis carinii
  – Aspergillus
  – *M. tuberculosis*
  – *M. avium*
  – CMV
Specimens
  – Sputum
  – Nasopharyngeal aspirate
  – Bronchoalveolar lavage
Diagnosis
  – Microscopy
  – Culture
  – Ag detection
  – PCR

Tumour or foreign body
  – Oral anaerobes
  – Staphylococci

Upper respiratory tract infections
Throat infection
This is a common condition in community practice, caused by both viruses, e.g. adenovirus, coxsackie virus and bacteria. *Streptococcus pyogenes* is the most common bacterial cause, but *Corynebacterium diphtheriae* infection should be considered with an appropriate travel history. *Neisseria gonorrhoeae* and *Candida* can cause pharyngitis.

Patients have fever and a painful infected throat that may have visible pus or exudate. Regional lymph nodes may be painful and enlarged. Diphtheria can cause a green–black necrotic pharyngeal 'membrane' associated with neck oedema in severe cases. Streptococcal infection may be complicated by peritonsillar abscess (quinsy), bacteraemia, rheumatic fever or nephritis.

If diphtheria is suspected, throat swabs should be taken and the laboratory alerted so that they will be inoculated onto appropriate media.

Symptomatic treatment is adequate for many infections but penicillin (penicillin V) or a macrolide (erythromycin) may be given. Ampicillin should be avoided as it may provoke a skin rash with Epstein–Barr virus infection. Tonsillectomy and adenoidectomy may reduce the number of infective episodes of pharyngitis or otitis media.

Otitis media and sinusitis

Infection occurs when sinuses or the middle ear are occluded by inflammation. Children under 7 years are especially prone to otitis media because the eustachian tube is short, narrow and nearly horizontal. The main infecting organisms are *Streptococcus pyogenes*, *Streptococcus pneumoniae*, *Haemophilus influenzae* and *Moraxella catarrhalis*.

Patients present with fever and local intense pain: small children may have difficulty in localizing the pain. In sinusitis, the pain is often worse with head movement and in the evening. Ear infection may be complicated by perforation, recurrent or chronic infection or the development of 'glue ear' (sterile mucus within the middle ear). Acute meningitis or mastoiditis may complicate severe infection.

Diagnosis is clinical; an auroscope reveals retrotympanic fluid levels or an inflamed tympanic membrane or a purulent discharge associated with perforation. Treatment depends on reducing mucosal swelling, promoting drainage of fluid and encouraging the recirculation of air. Appropriate antibiotic therapy has a role in this process.

Acute epiglottitis

This infection causes swelling of the epiglottis that may threaten the airway. *Haemophilus influenzae* type b was the most common cause until vaccination became available. Infection with *S. pyogenes* causes some cases, usually in adults. The child has a sore throat, high fever, and often stridor and drooling. Examination of the throat should be avoided as it may precipitate acute respiratory obstruction. Treatment is with parenteral third-generation cephalosporins. Tracheostomy may become necessary.

Lower respiratory tract infections

Lower respiratory tract infections (LRTIs) are an important cause of morbidity and mortality world-wide. They are the leading cause of death in children under the age of 5 years in developing countries.

Patients are predisposed to community-acquired pneumonia by factors including smoking, chronic obstructive pulmonary disease, diabetes mellitus, immunosuppressive therapy and HIV.

Many viruses cause primary viral pneumonia, e.g. influenza. Others cause damage to the lower respiratory tract permitting secondary bacterial pneumonia (see Chapter 28).

Clinical features

Patients have fever and a cough. Sputum may be purulent or blood-stained, although some infections, e.g. *Mycoplasma*, do not have a productive cough. Inflammation of the pleura causes sharp chest pain, worse on inspiration. Lower respiratory tract infections may also show signs of systemic infection, e.g. myalgia, malaise and weakness. In the elderly, mental confusion is common: specific symptoms and signs may be slight.

Infection, especially with *S. pneumoniae*, can be complicated by local spread to the pleura and pericardium, and by septicaemia and meningitis. *Staphylococcus aureus* infection can be complicated by lung cavitation and bronchiectasis after recovery.

Diagnosis

Sputum, not saliva, should be collected; physiotherapy may be helpful in obtaining a good quality specimen. Many patients are too ill to produce a sputum specimen, so bronchoalveolar lavage can be performed as an alternative and is valuable for diagnosis in immunocompromised patients.

Gram stain and direct microscopic examination may show sheets of Gram-positive pneumococci: direct immunofluorescence may confirm *Legionella*. Culture allows species identification and sensitivity testing. Antigen detection methods and PCR are available for *Chlamydia*, *Mycoplasma*, *Legionella*, *Coxiella* and *S. pneumoniae*.

Management and prevention

Appropriate antibiotic therapy should be commenced as soon as possible. Severe community-acquired pneumonia requires hospitalization, with intravenous antibiotics; milder infections can be treated orally. As β-lactamase resistance is common in *H. influenzae,* chronic obstructive pulmonary disease patients should be treated with an appropriate agent, e.g. co-amoxiclav or trimethoprim. Treatment of hospital-acquired pneumonia may require agents active against Enterobacteriaceae and *Pseudomonas*, e.g. ciprofloxacin or ceftazidime.

Supportive therapy, e.g. bedrest, oxygen, rehydration, physiotherapy and ventilation may be needed.

Infective exacerbations of cystic fibrosis require specialist management with changing regimens of antimicrobials and intensive postural drainage and physiotherapy.

44 Urinary and genital infections

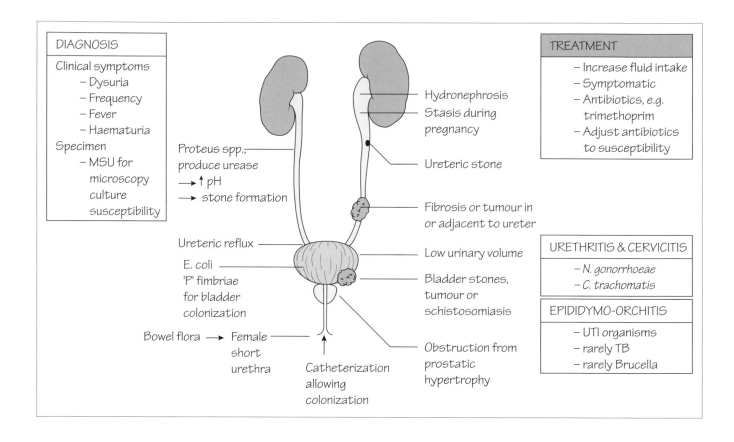

DIAGNOSIS

Clinical symptoms
- Dysuria
- Frequency
- Fever
- Haematuria

Specimen
- MSU for microscopy culture susceptibility

Proteus spp., produce urease
→ ↑ pH
→ stone formation

Ureteric reflux

E. coli 'P' fimbriae for bladder colonization

Bowel flora → Female short urethra

Catheterization allowing colonization

Hydronephrosis
Stasis during pregnancy

Ureteric stone

Fibrosis or tumour in or adjacent to ureter

Low urinary volume

Bladder stones, tumour or schistosomiasis

Obstruction from prostatic hypertrophy

TREATMENT
- Increase fluid intake
- Symptomatic
- Antibiotics, e.g. trimethoprim
- Adjust antibiotics to susceptibility

URETHRITIS & CERVICITIS
- N. gonorrhoeae
- C. trachomatis

EPIDIDYMO-ORCHITIS
- UTI organisms
- rarely TB
- rarely Brucella

Urinary tract infection

Anatomical considerations
Only the lower part of the urethra is usually colonized by bacteria; the flushing action of urinary flow protects against ascending infection. As the female urethra is short, urinary infection is more common in women.

Epidemiology and pathogenesis
Dehydration, obstruction, the disturbance of smooth urinary flow or the presence of a foreign body, e.g. stone or urinary catheter, may predispose an individual to urinary infection. Trauma during sexual intercourse may precipitate infection in women. Paediatric infection, especially in boys, is often associated with congenital abnormality, e.g. ureteric reflux or urethral valves.

The most commonly isolated pathogens are *Escherichia coli* and *Enterococcus* spp. *E. coli* uses fimbriae to adhere to the urinary epithelium, reducing the risk of being washed away. Infections caused by *Proteus* spp. are more likely in patients who have stones: *Proteus* spp. have urease activity that raises urinary pH, thus encouraging stone formation. *Staphylococcus saprophyticus* is a common isolate from sexually active females. Many different Gram-negative organisms colonize urinary catheters, often becoming invasive infections.

Clinical features
Lower urinary tract infections are characterized initially by urinary frequency, dysuria and suprapubic discomfort, fever may be absent. In pyelonephritis, fever, loin pain, renal angle tenderness and signs of septicaemia may be present. In children, the elderly and antenatal patients, urinary infection may be clinically silent. Recurrent infections can result in scarring and renal failure.

Laboratory diagnosis
The number of white blood cells and epithelial cells can be used to assess specimen quality and significance of isolates. Urine can be contaminated by perineal organisms. This is minimized by taking a mid-stream specimen and taking $> 10^5$ c.f.u. per ml of a single organism to differentiate infection from contamination ($< 10^5$ organisms per ml). This numerical approach is not appropriate for urine obtained from chronically catheterized patients where many organ-

isms may be present, or from suprapubic aspiration in infants with suspected infection where all isolates are potentially significant.

Susceptibility tests must be performed on all significant isolates.

Treatment and prevention

Antibiotic choice should be defined by susceptibility tests; empirical therapy should follow the known susceptibilities of urinary pathogens in that community. Most community-acquired infections respond to oral antibiotics, e.g. cefalexin, amoxicillin or trimethoprim. Should septicaemia develop, ciprofloxacin, cefotaxime or gentamicin may be required. Patients with recurrent urinary infection may require nocturnal prophylaxis, e.g. low-dose trimethoprim, nitrofurantoin or nalidixic acid, together with advice on ensuring adequate urine flow. Children with recurrent infections should be investigated and may require surgical correction of anatomical abnormalities. Significant bacteriuria in pregnant women should be treated, even if asymptomatic.

The risk of urinary tract infection is reduced by drinking enough fluids to ensure an adequate urinary flow. Anatomical obstructions to urine flow, such as stricture or stones, should be removed if possible.

Genital infection

Genital infection presents in many ways (see Table 43.1). Other sites may be involved, e.g. the throat and rectum in gonococcal infection. It may be followed by pelvic inflammatory disease, infertility, prostatitis, arthritis or bacteraemia.

Diagnosis

Urethral and cervical swabs should taken for both bacterial and viral culture. *Neisseria gonorrhoeae* may be isolated on New York City medium, and *Chlamydia* detected by tissue culture, EIA, PCR or ligase chain reaction. Viral culture may demonstrate herpes simplex (see Chapter 25). Serological diagnosis with EIA and a combination of traditional treponemal tests (see Chapter 22) may confirm syphilis. Direct microscopy may show evidence of *Candida* or *Trichomonas*.

Treatment

Chlamydia is responsive to a 2-week course of either tetracycline or a macrolide, e.g. erythromycin. Penicillins are still the treatment of choice for gonorrhoea if sensitive, although cephalosporins, quinolones or spectinomycin are usually required. Syphilis is treated with penicillin (see Chapter 22).

Table 44.1

Syndrome	Organisms
Genital ulcers	Herpes simplex *Chlamydia trachomatis* types L 1–4 *Haemophilus ducreyi* (see Chapter 16) *Treponema pallidum* (see Chapter 22) *Calymmatobacterium donovani*
Urethral discharge	*Neisseria gonorrhoeae* *C. trachomatis*
Pelvic inflammatory disease	*N. gonorrhoeae* *C. trachomatis* Mixed anaerobic infection
Vaginal discharge	*Candida albicans* *Trichomonas vaginalis* *Mobiluncus* spp. and others in non-specific vaginitis

Prevention

Prevention requires risk avoidance, e.g. monogamous relationships, or risk reduction, e.g. barrier contraceptive methods. Sexual contacts of cases are traced to treat asymptomatic disease and reduce transmission. There are no effective vaccines for sexually transmitted diseases.

Trichomonas vaginalis

This protozoan causes an itchy vaginal infection, which presents as a discharge with an offensive smell. Treatment is with metronidazole: treatment of sexual contacts may be necessary.

Non-specific vaginosis

A mixture of organisms, including *Mobiluncus* spp. and *Gardnerella vaginalis*, may result in an offensive discharge with a characteristic fishy smell when alkalinized. The diagnosis is based on the presence of epithelial cells heavily coated with bacteria in the discharge, and a positive amine test. Non-specific vaginosis is treated with metronidazole.

Epididymo-orchitis

Infection of the epididymis may arise from the urinary tract, or as part of a genital or systemic infection, e.g. brucellosis or tuberculosis. Patients present with acutely inflamed epididymis and testis that must be distinguished from testicular torsion. Diagnosis is made clinically and confirmed by the result of urinary cultures, or blood cultures.

45 Infections of the bones and joints

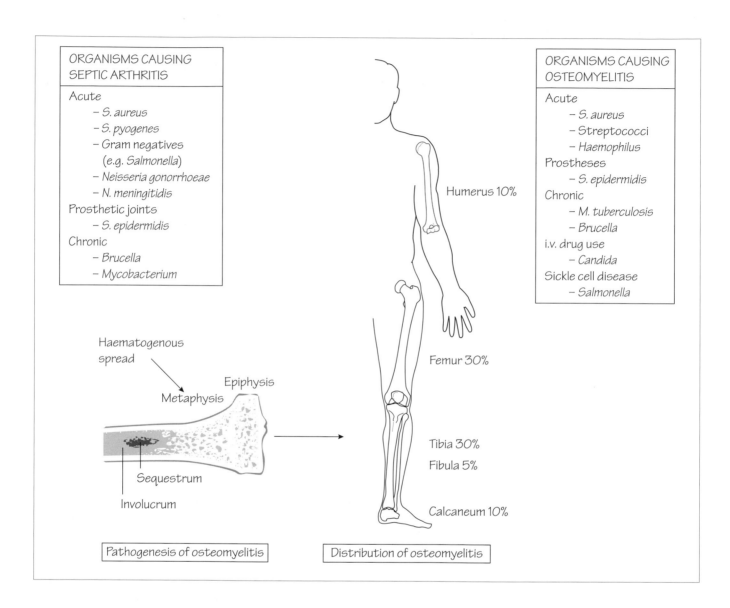

ORGANISMS CAUSING SEPTIC ARTHRITIS

Acute
- S. aureus
- S. pyogenes
- Gram negatives
  (e.g. Salmonella)
- Neisseria gonorrhoeae
- N. meningitidis

Prosthetic joints
- S. epidermidis

Chronic
- Brucella
- Mycobacterium

ORGANISMS CAUSING OSTEOMYELITIS

Acute
- S. aureus
- Streptococci
- Haemophilus

Prostheses
- S. epidermidis

Chronic
- M. tuberculosis
- Brucella

i.v. drug use
- Candida

Sickle cell disease
- Salmonella

Humerus 10%

Femur 30%

Tibia 30%

Fibula 5%

Calcaneum 10%

Haematogenous spread

Epiphysis

Metaphysis

Sequestrum

Involucrum

Pathogenesis of osteomyelitis

Distribution of osteomyelitis

Osteomyelitis
Osteomyelitis (infection of bone) may arise from haematogenous spread, by direct extension from an infected joint, or following trauma, surgery or instrumentation. The formation of pus precipitates ischaemia and necrosis; the central area of dead bone is known as the sequestrum. New bone (the involucrum) may form around the infection site. In children, the metaphysis of the long bones (femur, tibia and humerus) are most often involved. In addition to these sites, infection of the spine is common in adults.

Staphylococcus aureus accounts for 90% of infections: rarer causes include *Streptococcus pyogenes* (4%), *Haemophilus influenzae* (4%), *Escherichia coli*, *Salmonella* spp., *Mycobacterium tuberculosis* and *Brucella*. Patients with sickle cell disease are especially prone to *Salmonella* infection.

Clinical features
Patients present with fever and poorly localized pain. Young children may stop moving the affected limb (pseudoparalysis). As infection progresses, soft-tissue swelling may occur which may be followed by sinus formation. Pathological fractures may develop if diagnosis and treatment are delayed.

Diagnosis

Radiological changes do not develop until late in the course of infection when demineralization has occurred. Isotope scans may be helpful but do not distinguish infection from other inflammatory conditions. Blood cultures are often negative early in the course. Pus from bone via needle or open biopsy allows culture for identification and susceptibility testing.

Management

Drainage and excision of the sequestrum is an important part of the management. Concurrent antibiotic therapy, e.g. flucloxacillin and fusidic acid, should be started empirically pending culture results. Other agents, e.g. ciprofloxacin, may be required if, for example, *Salmonella* is isolated. Treatment lasts for 6 weeks or until there is evidence that inflammation has disappeared and the bone has healed.

Chronic osteomyelitis

Chronicity may arise from inadequately treated acute infection or secondary to surgery or fracture. *Staphylococcus aureus* is implicated in 50% of cases; the remainder is associated with Gram-negative pathogens (*Pseudomonas*, *Proteus*, and *E. coli*). Ongoing pain, swelling and deformity, with a chronically discharging sinus, are the main clinical features. A diagnosis by culture is essential; specimens should be taken under aseptic conditions. A prolonged course of appropriate antibiotics should accompany appropriate surgery.

Suppurative arthritis

Suppurative arthritis usually arises from a bacteraemia; 95% of cases are caused by *S. aureus* and *Streptococcus pyogenes*. Other causes include Enterobacteriaceae, *Neisseria gonorrhoeae*, *H. influenzae*, *Salmonella* spp., *Brucella* spp., *Borrelia burgdorferi*, *Pasteurella* and *M. tuberculosis*. The large joints, e.g. the knee, are most commonly infected, but infection of the shoulder, hip, ankle, elbow and wrist joints may also occur. Prosthetic joints may become infected with skin contaminants (usually *S. aureus* or *Staphylococcus epidermidis*) at the time of operation, or from haematogenous spread. The original source will dictate the causative pathogen.

Clinical features

In children, the onset may be abrupt, with fever, pain and swelling of the joint associated with reduced movement. In adults, the onset may be insidious: a history of recent urinary infection or salmonellosis may be reported. Other associated signs include cellulitis or specific rashes, e.g. gonococcal skin rash.

Septic arthritis must be differentiated from acute rheumatoid arthritis, osteoarthritis, gout, pseudogout or reactive arthritis. A diagnostic tap will yield cloudy fluid. A Gram stain and white blood cell count may suggest infection that can be confirmed by culture within 48 h. Bone marrow should be obtained when brucellosis is suspected.

Intravenous antibiotics, appropriate to the infecting organisms isolated or suspected, should be commenced and oral therapy continued for up to 6 weeks. Aspiration and irrigation of the joint may be helpful in severe cases by reducing inflammatory damage.

Viral arthritis

Some viruses are associated with arthritis, e.g. rubella, mumps and hepatitis B. Rubella-related arthritis is more common in females and develops a few days after the rash. Several of the alphaviruses cause severe bone and joint symptoms. Arthritis, caused by an immune response to the pathogen, can follow recovery, e.g. after meningococcal disease, *Chlamydia* or *Shigella* infection. This last infection can be associated with uveitis and is known as Reiter's syndrome. Alternatively, haematogenous spread of the infecting organisms may give a septic arthritis.

Prosthetic joint infections

Prosthetic joints may become infected at the time of operation or as a result of haematogenous spread. Organisms are often of low virulence, such as *S. epidermidis*. Infection with *S. aureus*, especially if methicillin-resistant (MRSA), can have serious consequences. Treatment is with intravenous antibiotics, depending on the susceptibility of the infecting organisms. Infection often results in loss of the prosthesis: it is important to prevent infection by effective control measures in the ward and theatre. Patients undergoing prosthetic joint surgery should receive antibiotic prophylaxis with an agent active against *S. aureus*.

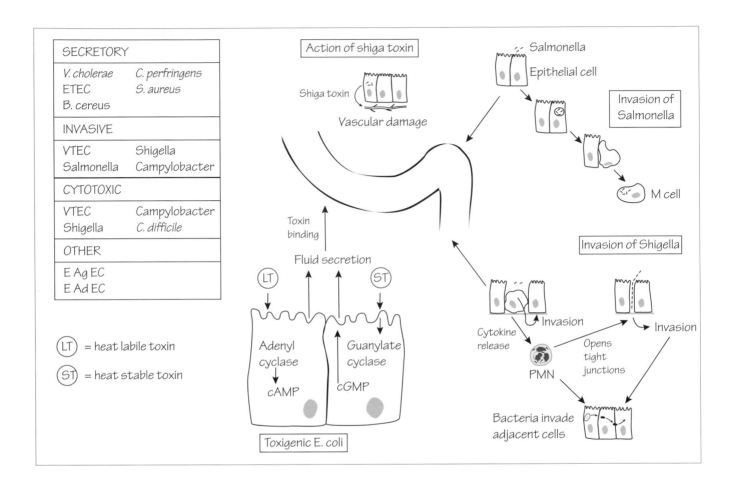

Infectious diarrhoea is a common condition resulting in considerable economic loss from work absence. It is one of the most important causes of death for children under 5. The gut is protected by gastric acid, bile salts, the mucosal immune system and the inhibitory substances produced by the normal flora.

Organisms are transmitted by hands and fomites (faecal–oral route), by food or by water. The infective dose can be as few as 10 organisms (*Shigella*). Some foods, e.g. milk, or drugs, e.g. H_2 antagonists and proton pump inhibitors, may reduce the protective effects of gastric acid.

Food is an important vehicle of infectious diarrhoea. Bacteria enter the food chain from animal infections, from poor hygiene during slaughter, and during butchering. Hens that are chronically colonized with *Salmonella* produce eggs that may be contaminated. Improper cooking and storage may allow the multiplication of bacteria (see below).

Transmission of diarrhoeal disease is also facilitated where there is poor sanitation, e.g. general poverty, war or refugee crises. In these situations, infection spreads rapidly through the community, causing significant mortality. Cholera is capable of spreading world-wide (a pandemic).

Traveller's diarrhoea usually develops within 72 h of arrival in a new country: Latin America, Africa and Asia are the regions with the highest risk. Patients experience two to four watery bowel motions daily; blood and mucus are typically absent. The major organisms implicated are enterotoxigenic and enteroadherent *Escherichia coli*. Treatment is with fluid replacement and antibiotics, including co-trimoxazole or ciprofloxacin.

Pathogenesis

Infectious diarrhoea causes symptoms by a number of mechanisms, e.g. toxic deregulation of intestinal cells causing fluid secretion or invasion of the intestinal wall with destruction of the cells (see diagram). Secretory diarrhoea produces infrequent large-volume stools as the absorptive capacity of the colon is overwhelmed. In dysenteric illness (*Shigella*), inflammation of the colon causes a loss in the capacitance, resulting in frequent stools that are often blood-stained.

Clinical features

Although diarrhoea may be defined as an increase in frequency of bowel action, it is a very subjective symptom. There may be many small stools (typical of large bowel infection), or infrequent large stools (small intestine infection). Stools may be blood-stained when there is destruction of the intestinal mucosa, or have a fatty consistency and offensive smell if malabsorption is present.

Dehydration and electrolyte imbalance may develop rapidly with potentially fatal results, e.g. cholera. Crampy abdominal pain may accompany diarrhoea, e.g. *Campylobacter* and *Shigella* infections; this may mimic acute abdominal conditions, e.g. appendicitis. Fever is not always present in diarrhoeal disease.

Septicaemia may develop in some cases of salmonellosis but is rare in other diarrhoeal diseases. Self-limiting bacteraemia is common in *Campylobacter* infection. *Escherichia coli* O157 infection can produce a haemorrhagic colitis that is later complicated by renal failure and the haemolytic-uraemic syndrome. Secondary lactose intolerance resulting in continuing diarrhoea is caused by loss of intestinal lactase. It usually lasts a few weeks before resolving spontaneously. Patients with immunodeficiency may have difficulty eradicating intestinal infections: IgA deficiency, *Giardia*; T cell, *Salmonella* and *Cryptosporidium* (see Chapter 48). Diarrhoea from protozoa or viruses are discussed in more detail in Chapters 34 and 29, respectively.

Diagnosis

Stool should be routinely examined microscopically for intestinal protozoa (e.g. *Giardia lamblia*). Ziehl–Nielsen's stain can be used to detect microsporidia and *Cryptosporidium parvum* (see Chapter 34).

Selective media must be used to culture bacterial pathogens so that the growth of non-pathogenic commensals is suppressed, e.g. sorbitol MacConkey for verotoxic *E. coli* (O157). Media can be made selective for *Campylobacter* by incorporating antibiotics and/or by incubating the plates at 43°C. If cholera is suspected, stools are inoculated into alkaline peptone water (high pH allows *V. cholerae* to grow preferentially): it can then be subcultured onto special selective medium containing bile salts and a high pH.

Organisms may be typed for epidemiological purposes, allowing investigators to follow the spread of pathogens in a community, e.g. cholera, or to investigate outbreaks of food-borne disease so that preventive measures can be implemented. When organisms have only one serotype, e.g.

Shigella sonnei, further typing (colicine typing) is required to confirm an outbreak.

The presence of viruses in stool can be demonstrated directly by electron microscopy, culture or EIA (see Chapter 29). Toxin may be detected in stool samples, e.g. *C. difficile* toxin.

Management

The management of diarrhoeal disease is based on adequate fluid replacement and restoration of electrolyte imbalances. Despite the outflow found in secretory diarrhoea, fluid absorption still occurs. Simple oral rehydration solutions consist of 150–155 mmol sodium/L, 200–220 mmol/L glucose and can be life-saving. Intravenous fluid replacement is rarely necessary. Antimotility drugs are of no benefit and may be dangerous, especially in small children. Cholera and severe fluid diarrhoea may benefit from oral antibiotics, such as tetracycline or ciprofloxacin, which may shorten the duration of symptoms. Patients with severe dysentery and salmonellosis should be treated with ciprofloxacin or co-trimoxazole. Chronic infective diarrhoea is often caused by *Giardia* or amoebic infection, and will require treatment with metronidazole. Renal failure associated with haemolytic-uraemic syndrome following *E. coli* O157 requires specialist management.

Prevention

There are several candidate vaccines for cholera, e.g. live vaccines attenuated by genetic means, but none are yet licensed. There are several vaccines for typhoid; none is available for salmonellosis.

Provision of a safe water supply, uncontaminated by human or animal faeces, is central to the prevention of diarrhoeal disease. Animal husbandry and slaughter methods should be designed to prevent the introduction of animal intestinal pathogens into the human food chain. Food must be cooked to a sufficiently high temperature to kill pathogens and, if not eaten immediately, refrigerated at a low enough temperature to prevent bacterial multiplication.

Cooked food should be physically separated from uncooked to prevent cross-contamination. This is especially true in institutional cooking, e.g. hospitals and restaurants, where many may become infected following a single failure of hygiene.

Traveller's diarrhoea can be prevented by careful choice of food while travelling.

47 Zoonoses

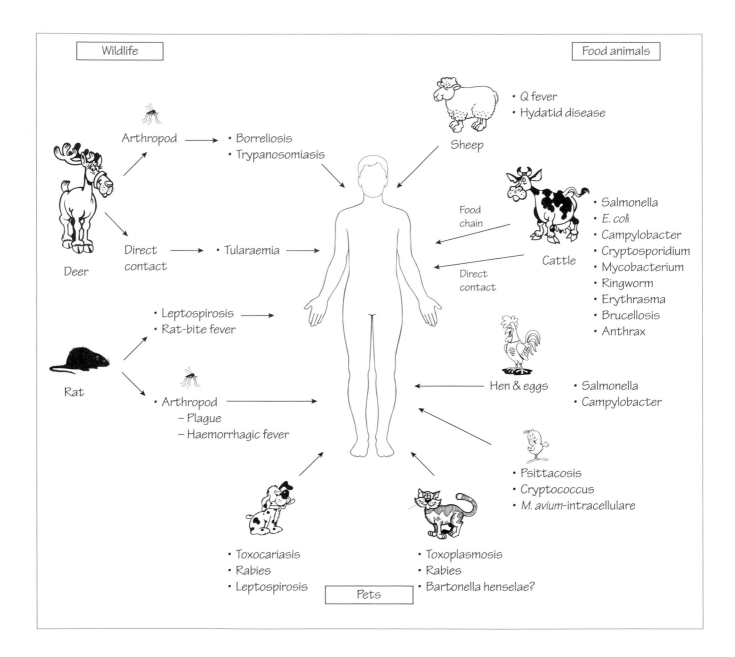

A zoonosis is an infection acquired from an animal source. Infections are acquired when humans enter the environment where the natural life-cycle occurs, e.g. camping. Transmission can occur via vectors, e.g. mosquitoes (Japanese B encephalitis). Alternatively, farming may expose the workers to infections from livestock. Pets are an important source.

Viral zoonoses

More than 100 animal viral pathogens are implicated in human disease. For example, herpes simiae, a monkey pathogen, can be transmitted to a human handler by monkey bite. It usually causes a severe encephalitis with 75% mortality. Early treatment with aciclovir may be beneficial. Other viral zoonoses are discussed in Chapter 31.

Rat bite fever

Rat bite fever is caused by a Gram-negative spiral bacterium, commensal in the nasopharynx of rats and transmitted to humans by the bite of a rat. Following a 2-week incubation period, an inflammatory reaction is found at the site of the bite with lymphangitis and regional lymphadenopathy. There is a generalized maculopapular rash, together with fever, headache and malaise. Endocarditis is the most serious complication. Spontaneous recovery may occur within 2 months.

Diagnosis relies on visualizing the organism in blood or lymph node using Giemsa stain as it has not yet been cultivated *in vitro*. Treatment is usually with penicillin.

Anthrax
See Chapter 12

Plague
Caused by *Yersinia pestis*, the infection is endemic in rodents in remote rural areas. Rarely, epidemics may develop which may spread world-wide, e.g. the Black Death. The organism is transmitted between rats, and to humans, by the rat flea, *Xenopsylla cheopis*.

The incubation period is short: the disease has an abrupt onset characterized by fevers and toxaemia. The regional lymph glands draining the site of the bite become greatly enlarged (buboes) and septicaemia is accompanied by generalized haemorrhage. Pneumonic plague is a rapidly fatal pneumonitis that can be transmitted by the respiratory route.

Plague is diagnosed clinically in areas where it is endemic. A Gram stain of lymph gland aspirate or culture of blood will demonstrate the organism. Treatment is with tetracycline, chloramphenicol or aminoglycosides. The mortality rate of pneumonic plague is high.

Borreliosis
Borreliosis is transmitted from rodents or deer by ticks (open forest habitat, e.g. the New Forest) or by lice (see Chapter 22).

Toxoplasmosis
The cat is the definitive host of toxoplasmosis but the organism infects a wide range of animals, including sheep, cattle and humans. Infection is acquired by ingestion of oocysts from infected cat faeces or from tissue cysts in infected meat, e.g. undercooked beef (see Chapter 38).

Dermatophytes
Dermatophytes that are natural pathogens of animals can spread to the human population by direct contact (see Chapter 33).

Toxocariasis
Toxocara canis is an ascarid parasite of dogs. The parasite eggs are excreted in the faeces of infected dogs and mature in the soil. Human ingestion occurs when food is contaminated by the soil, or when personal hygiene is poor, e.g. hand washing. The larval stages hatch in the intestine, invade the host and migrate to the liver and lungs. They are unable to develop into adults but migrate throughout the body causing fever, hepatosplenomegaly, lymphadenopathy and wheeze. If the larva migrate into the eye, sight may be permanently damaged by local inflammatory response of the retina. The diagnosis is made serologically using a specific EIA. The disease is usually self-limiting but, if symptoms are severe, treatment with albendazole may be beneficial. Ocular lesions should be treated first with steroids to diminish the inflammatory response; the role of antihelminthic treatment is less certain.

Cat scratch disease
Ten days following a cat scratch or bite, a papular lesion caused by *Bartonella henselae* may develop at the site. It is associated with regional lymphadenopathy. The symptoms resolve slowly over a period of 2 months, but a more chronic course may ensue. Cat scratch disease can be complicated by disseminated infection: this is more common in immunocompromised individuals. Diagnosis is usually made clinically but can be confirmed serologically by immunofluorescence or EIA. The organism is fastidious, requiring a prolonged incubation period. PCR may also be used diagnostically. Treatment with erythromycin, tetracyclines or rifampicin may be beneficial.

Hydatid disease
Two species of parasite are responsible for human hydatid disease: *Echinococcus granulosus* and *E. multilocularis*. Dogs are the definitive host, harbouring the tapeworm stage; the eggs are passed in the faeces. The eggs are ingested by the intermediate hosts, e.g. sheep or rodents, and multiple cysts develop in the liver and lungs. The cycle is complete when dogs eat infected tissues. Humans are accidental hosts. The disease is common in sheep-farming areas. *E. multilocularis* is found in foxes, wolves and dogs; rodents act as the intermediate hosts.

Pathogenesis and clinical features
Cysts act as space-occupying lesions in the liver, lungs, abdominal cavity or central nervous system and are responsible for the symptoms and signs of disease. The cysts of *E. multilocularis* lack a definite cyst wall and may ramify widely in the tissue.

Diagnosis
Cysts may be demonstrated by ultrasound or CT. EIA for both antibody and antigen is available.

Treatment
If possible, hydatid cysts should be surgically removed. Albendazole is given to kill the germinal layer of the cyst, and praziquantel to reduce the viability of protoscolices. If the cysts rupture and viable protoscolices escape, multiple cysts can form in the abdomen or sudden antigen release can provoke acute anaphylaxis.

48 Infections in immunocompromised patients

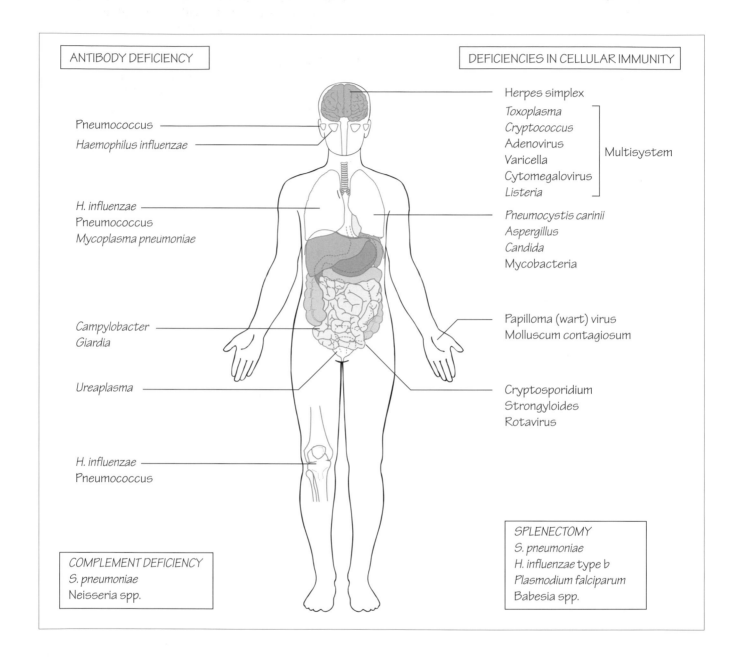

ANTIBODY DEFICIENCY

DEFICIENCIES IN CELLULAR IMMUNITY

Herpes simplex

Toxoplasma
Cryptococcus
Adenovirus   Multisystem
Varicella
Cytomegalovirus
Listeria

Pneumococcus
Haemophilus influenzae

Pneumocystis carinii
Aspergillus
Candida
Mycobacteria

H. influenzae
Pneumococcus
Mycoplasma pneumoniae

Papilloma (wart) virus
Molluscum contagiosum

Campylobacter
Giardia

Ureaplasma

Cryptosporidium
Strongyloides
Rotavirus

H. influenzae
Pneumococcus

SPLENECTOMY
S. pneumoniae
H. influenzae type b
Plasmodium falciparum
Babesia spp.

COMPLEMENT DEFICIENCY
S. pneumoniae
Neisseria spp.

Medical treatment or hereditary conditions may allow organisms with low virulence to cause infection. The pathogenesis is usually multifactorial, e.g. patients undergoing bone marrow transplantation are neutropenic and have a reduced resistance to bacterial infection; intravenous cannulation provides a route for *S. epidermidis* infection.

Disorders of the innate immune system are found when medical treatment breaches the physical barriers (see Chapter 7). Infections specifically associated with AIDS are discussed in Chapter 39.

Neutropenia

Granulocytopenia most often arises as a result of acute leukaemia or its treatment. The risk of infection depends on both the duration and severity of the neutropenia. Bacteraemia occurs in between 40 and 70% of neutropenic patients. The Enterobacteriaceae and *Pseudomonas* spp. are the most common Gram-negative bacilli isolated. These bacteria invade following damage to the intestine by antineoplastic agents or irradiation. Gram-positive organisms (*S. epidermidis*, *S. mitis* and *S. oralis*, *Enterococcus* spp., *S.*

aureus and *Corynebacterium jeikeium*) are increasingly important causes of sepsis.

Although antibiotic therapy may predispose to colonization by *Candida albicans*, fungal infection may occur *de novo* in the neutropenic patient. Increasingly, yeasts, e.g. *Candida krusei* (naturally resistant to antifungal therapy), *Aspergillus* spp. (causing invasive disease), and infections with *Fusarium* spp., *Pseudallechera boydi* and *Trichosporon beigeli* are being reported.

Treatment of fever in neutropenic patients
Empirical therapy includes carbapenem or ceftazidime and amikacin. If fever is unresolved, a glycopeptide can be added. Later, if fever still persists, fungal pathogens are more likely and amphotericin can be added.

Prevention of infection
The risk of infection in neutropenic patients is reduced when the patient is nursed in a side room and supplied with sterilized water and food. Sterile procedures, e.g. thorough hand-washing and latex gloves, should be employed: attendants should also wear gowns and masks. Room air is filtered to remove fungal spores.

Antibiotic prophylaxis using 4-fluoroquinolones, targets the facultative anaerobes of the gut. This preserves the anaerobic flora, preventing recolonization with potentially pathogenic Gram-negative bacilli. Oral nystatin, alone or in combination with oral amphotericin, reduces the incidence of fungal infection. Fluconazole or itraconazole may also be useful.

T cell deficiency
This is an increasingly common problem following HIV infection, cancer chemotherapy, corticosteroid therapy or organ transplantation. Congenital T cell deficiencies are rare but may be purely linked to T cell function or combined with a hypogammaglobulinaemia.

Pathogens
These are mainly those microorganisms which, in the human host, have an intracellular location, e.g.
• *T. gondii*, *Strongyloides stercoralis*;
• *M. tuberculosis*, *M. avium-intracellulare*;
• *Listeria monocytogenes*, *Cryptococcus neoformans*, *Pneumocystis carinii*;
• herpes simplex, cytomegalovirus, varicella-zoster virus and measles.
Measles infection, complicated by giant cell pneumonia and encephalitis, can be life-threatening.

Diagnosis
Specific infections should be investigated appropriately (see relevant chapters). All patients should have at least two blood cultures taken from different sites.

Hypogammaglobulinaemia
X-linked agammaglobulinaemia patients are at increased risk of infection for the first 6 months of life; the common, variable immunodeficiency patients are at increased risk throughout life. Functional hypogammaglobulinaemia develops in patients with multiple myeloma.

Patients suffer recurrent respiratory tract infections with *S. pneumoniae* and non-capsulate *H. influenzae* leading to bronchiectasis. *Giardia*, *Cryptosporidium* and *Campylobacter* infections may be more persistent. Intravenous immunoglobulin reduces recurrent infection.

Complement deficiency
Hereditary complement deficiencies are rare. Deficiency in the later components of the complement cascade (C7–9) results in an inability to lyse Gram-negative bacteria, and are susceptible to recurrent *Neisseria* infection. Deficiency of the alternative complement pathway leads to serious *S. pneumoniae* infections, including meningitis. Acquired complement deficiency occurs in systemic lupus erythematosus.

Postsplenectomy
The incidence of serious sepsis following splenectomy is about 1% per year; the rate is higher in infants and children. Highest mortality is associated with splenectomy for lymphoma and thalassaemia. Although the risk of sepsis diminishes with time, it never disappears. Patients with sickle cell disease have functional asplenia.

Streptococcus pneumoniae is responsible for approximately two-thirds of infections in most series; other important bacteria are *H. influenzae* and *Escherichia coli*. Malaria may run a fulminant course. Splenectomy predisposes to *Capnocytophaga canimorsis* infection, usually arising after a dog bite.

Prevention
Vaccination against pneumococcal disease should be offered to all splenectomized patients. Improved responses may be obtained with protein conjugate vaccines currently under trial.

Low-dose oral antibiotic prophylaxis with penicillin V should also be offered. Patients must be aware of the need to consult their doctor at the onset of any fever, or be instructed in the use of antibiotics, prescribed in advance to avoid delay.

49 Ocular infections

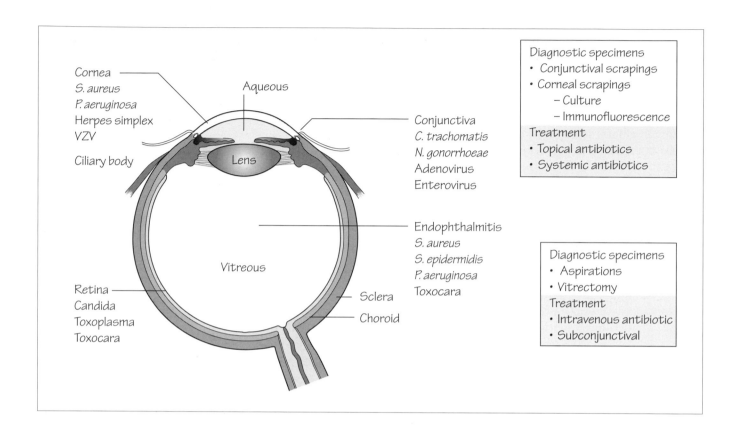

Diagnostic specimens
- Conjunctival scrapings
- Corneal scrapings
  - Culture
  - Immunofluorescence

Treatment
- Topical antibiotics
- Systemic antibiotics

Diagnostic specimens
- Aspirations
- Vitrectomy

Treatment
- Intravenous antibiotic
- Subconjunctival

Bacterial conjunctivitis

Bacterial conjunctivitis is a common condition caused by *Staphylococcus aureus*, *Haemophilus influenzae*, *Streptococcus pneumoniae* or *Moraxella* spp. Neonatal conjunctivitis may be caused by *Neisseria gonorrhoeae*, *Chlamydia trachomatis*, *Escherichia coli*, *S. aureus* and *H. influenzae* and is acquired from infection in the mother's genital tract. Infection with *Pseudomonas aeruginosa* can be acquired in hospital if ocular equipment or drops are not adequately sterilized or for single use. It produces a rapidly progressive infection that can result in ocular perforation and loss of vision. Irrespective of its aetiology, bacterial conjunctivitis presents with hyperaemic red conjunctiva and a profuse mucopurulent discharge. Conjunctival swabs and corneal scrapings are submitted for laboratory examination. The diagnosis is confirmed by bacterial culture; *Chlamydia trachomatis* antigen is confirmed by EIA or PCR. Treatment is by local antibiotics, including fusidic acid, tetracycline or chloramphenicol.

Adenovirus infection

Serotypes 7, 3, 10, 4 and 8 are the most common serotypes associated with ocular infection. Infection causes a purulent conjunctivitis, with enlargement of the ipsilateral periauricular lymph node. Half of patients with corneal involvement develop punctate keratitis followed by subepithelial inflammatory infiltration. Anterior uveitis and conjunctival haemorrhages may develop. Treatment is symptomatic, with antibacterial agents being used if there is evidence of secondary bacterial infection. Topical steroids should be avoided.

Varicella-zoster virus

The ophthalmic dermatome of the fifth cranial nerve is involved in approximately 10% of recurrent varicella-zoster virus (VZV) infection (shingles). Ocular involvement, associated with lesions present on the skin of the tip of the nose, includes anterior uveitis, keratitis, ocular perforation or retinal involvement. Chronic disease occurs in about a quarter of patients. The condition is very painful and may continue after healing of the rash (postherpetic neuralgia). Antiviral agents, e.g. aciclovir, should be used early in the infection and may prevent complications. Severe inflammation may benefit from topical steroids. A live attenuated vaccine is available to prevent primary infection.

Herpes simplex

Ocular infection with herpes simplex is the most common infectious cause of blindness in developed countries. Typically, it presents with ulcerative blepharitis, follicular conjunctivitis and regional lymphadenopathy. Most patients have corneal involvement. Relapses occur approximately every 4 years. Initially, the dendritic ulcer is the marker of infection, but the later clinical picture is dominated by inflammation in deeper tissues, keratitis, corneal oedema and opacity. Primary infection and early relapses are treated with topical aciclovir. Inappropriate use of steroids worsens the keratitis. Progressive scarring, following repeated attacks, leads to corneal opacity and is one of the most common indications for corneal grafting.

Ocular manifestations of AIDS

'Cotton wool spots' are a common retinal manifestation of HIV infection. They follow infarction of the retinal nerve fibre layer, and may result in poor colour sensitivity and perception. Late in the course of HIV infection, when the CD4 count has fallen to below $<0.05 \times 10^9$ per litre, ocular infection with cytomegalovirus may develop in up to one-third of patients. This causes a slowly progressive retinitis characterized by necrosis, and is an important cause of blindness in this patient group. The syndrome is difficult to differentiate from ocular toxoplasmosis or syphilitic retinitis. Initially, treatment with antiviral agents, e.g. ganciclovir, is given intravenously to control the disease; weekly maintenance therapy is required to prevent relapse once control is established.

Trachoma

This is a chronic keratoconjunctivitis caused by infection with *Chlamydia trachomatis*. It was once endemic throughout the world but is now largely confined to the tropics, where poor social conditions make transmission more easy and poverty precludes adequate medical treatment. Symptoms develop 3–10 days after infection, with lacrimation, mucopurulent discharge, conjunctival injection and follicular hypertrophy. Treatment is with oral macrolides, such as azithromycin.

Endophthalmitis

Endophthalmitis develops after ocular operation, following trauma and direct inoculation of a foreign body and as a complication of systemic infection. Early postoperative infections are usually with *S. aureus*, *Staphylococcus epidermidis*, streptococci and Gram-negative bacilli. Late postoperative infections are with more indolent bacteria derived from the skin or acute infections caused by streptococci or *H. influenzae*. Post-traumatic infections are with *S. epidermidis*, *Bacillus* and streptococci. Endogenous infections are most often with *Candida*, streptococci and enteric Gram-negative bacilli. Rarely, endophthalmitis is caused by the nematode *Toxocara canis* (see Chapter 47).

Diagnosis is achieved by taking vitreous aspiration or vitrectomy specimens. Antibiotics should be given intravenously and by subconjunctival injection.

Onchocerciasis

Onchocerciasis is one of the most important causes of blindness in the world. It is caused by the filarial parasite *Onchocerca volvulus*. Heavy infection causes inflammatory lesions in the eye, which result in blindness. This parasite is discussed in more detail in Chapter 37.

50 Infections of the skin and soft tissue

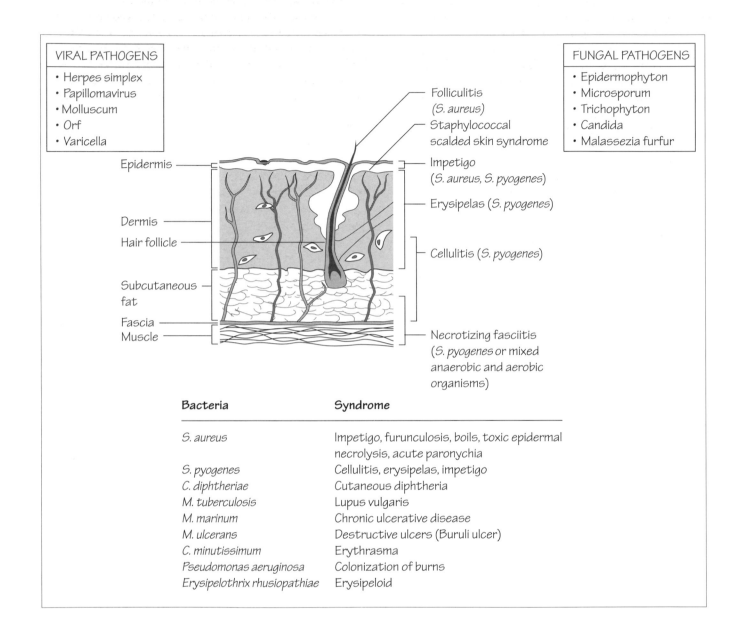

VIRAL PATHOGENS
- Herpes simplex
- Papillomavirus
- Molluscum
- Orf
- Varicella

FUNGAL PATHOGENS
- Epidermophyton
- Microsporum
- Trichophyton
- Candida
- Malassezia furfur

Folliculitis (*S. aureus*)

Staphylococcal scalded skin syndrome

Impetigo (*S. aureus, S. pyogenes*)

Erysipelas (*S. pyogenes*)

Cellulitis (*S. pyogenes*)

Necrotizing fasciitis (*S. pyogenes* or mixed anaerobic and aerobic organisms)

Epidermis

Dermis

Hair follicle

Subcutaneous fat

Fascia

Muscle

Bacteria	Syndrome
S. aureus	Impetigo, furunculosis, boils, toxic epidermal necrolysis, acute paronychia
S. pyogenes	Cellulitis, erysipelas, impetigo
C. diphtheriae	Cutaneous diphtheria
M. tuberculosis	Lupus vulgaris
M. marinum	Chronic ulcerative disease
M. ulcerans	Destructive ulcers (Buruli ulcer)
C. minutissimum	Erythrasma
Pseudomonas aeruginosa	Colonization of burns
Erysipelothrix rhusiopathiae	Erysipeloid

Bacterial

Skin infections spread rapidly by contact, especially in enclosed populations or those with poor sanitation. A wide range of organisms infects the skin (see diagram): *Staphylococcus aureus* and *Streptococcus pyogenes* are most commonly implicated.

Cellulitis effects all layers of the skin and can be caused by *S. pyogenes*, *S. aureus*, *Pasteurella multocida*, marine vibrios or Gram-negative bacilli. Organisms invade via skin abrasions, insect bites or wounds. Empirical flucloxacillin should be given until culture results are available. Severe disease should be trated with intravenous antibiotics, including benzylpenicillin and flucloxacillin.

Necrotizing fasciitis is a rapidly progressive infection that spreads to involve skin and subcutaneous layers. Mixed aerobic and anaerobic infection or pure *S. pyogenes* infection may be responsible. It progresses rapidly and leads to death in a very short time. Effective treatment depends on adequate surgical resection of infected tissue, supplemented with benzylpenicillin, a third-generation cephalosporin and metronidazole.

Erythrasma is a superficial infection of the flexures

caused by *Corynebacterium minutissimum*. Its lesions fluoresce under Wood's light. The organism may be cultured and treatment is with erythromycin or tetracycline.

Erysipelas is a demarcated streptococcal infection confined to the dermis, usually found on the face or shins, that feels hot and red. There is a modest increase in the peripheral white blood cells and fever may be present. Treatment with oral amoxicillin or flucloxacillin is usually effective but intravenous therapy may be required.

Erysipeloid, a dull red lesion, is a zoonosis derived from inoculation injuries, usually in pig-handlers. Treatment is with oral penicillin or tetracycline.

Burns are very susceptible to colonization by bacteria; in addition to *Pseudomonas aeruginosa*, *S. aureus* and *S. pyogenes* are implicated. Colonization can lead to the loss of skin grafts and to secondary bacteraemia.

Paronychia

This is a common infection in community practice. The cuticle is damaged, allowing invasion with organisms such as *S. aureus*. There is pain and swelling, followed by a small abscess. The abscess may be lanced and antibiotics given, e.g. flucloxacillin.

Manifestations of systemic infections

The skin is a large organ that can act as a window onto systemic infection. Examples include the petechial rash of meningococcal septicaemia heralding overwhelming sepsis. Patients with *Pseudomonas* septicaemia may have a gangrenous skin lesion, ecthyma gangrenosum. More subtle are the changes associated with endocarditis such as splinter haemorrhages. Staphylococcal septicaemia may have evidence of skin infarctions. Many viruses cause lesions in the skin, e.g. as part of a systemic infection (chickenpox and measles). In herpes simplex infection, the skin is the primary site of infection (see Chapter 24).

Warts

Human papillomaviruses infect skin cells causing increased skin replication and giving rise to a wart. Papular, macular or mosaic variations can occur; verrucae (plantar warts) are found on soles of the feet. The virus is transmitted by direct contact, particularly under wet conditions, e.g. around swimming pools. Genital warts (condylomata acuminata) may be transmitted sexually. The diagnosis of warts is usually made clinically; virus in condylomata acuminata can be detected by immunofluorescence and PCR-based amplification techniques.

Papillomaviruses are associated with malignancy: cervical (HPV-16 and HPV-18); laryngeal (HPV-6 and HPV-11). If warts are noted on routine cervical cytology, this is an indication for more frequent review.

Except in the immunocompromised, warts are self-limiting and spontaneously resolve without scarring. Topical keratolytic agents, e.g. salacylic acid, are widely available over the counter for self-application. Genital warts may respond to the application of podophyllum by trained staff. Cryotherapy may hasten their departure. Cautery is no longer recommended.

Several poxviruses infect the skin and cause characteristic lesions, e.g. molluscum contagiosum, orf. These are discussed in Chapter 26.

Dermatophytes
Clinical features

Dermatophyte infection (ringworm) may present as itchy red scaly patch-like lesions which spread outwards leaving a pale, healed centre. Chronic nail infection produces discoloration and thickening whereas scalp infection is often associated with hair loss and scarring. Clinical diagnostic labels are based on the site of infection, e.g. tinea capitis (head and scalp), tinea corporis (trunk lesion). The causative organism may vary.

Laboratory diagnosis

Infection of skin and hair by some species may demonstrate a characteristic fluorescence when examined under Wood's light.

Skin scrapings, nail clippings and hair samples should be sent dry to the laboratory. Typical branching hyphal elements may be demonstrated by potassium hydroxide. Dermatophytes take up to 4 weeks at 30°C to grow on Sabouraud's dextrose agar.

Identification is based on colonial morphology, microscopic appearance (lactophenol blue mount), physiological and biochemical tests.

Treatment

Dermatophyte infections may be treated topically with imidazoles, e.g. miconazole, clotrimazole, tioconazole or amorolfine. Some infections require oral terbinafine for several weeks.

Glossary

Accidental host: Humans can be accidental hosts of animal pathogens when they intrude on the normal cycle of transmission by contact with animals, e.g. hydatid disease.

Aerobe: An organism that grows in the presence of oxygen.

Aminoglycoside: A group of antibiotics, including gentamicin, which inhibit bacterial protein synthesis.

Amoeba: A single cell microorganism that moves by the use of pseudopodia. *Entamoeba histolytica* is the most important amoeba pathogenic for humans.

Anaerobe: An organism which grows in the absence of oxygen. Some may tolerate the presence of oxygen (aerotolerant), others be killed rapidly by any oxygen (obligate anaerobes). Organisms that grow equally well in either environment are called facultative anaerobes.

Antibiotic: A drug which inhibits the growth of a microorganism. This may be a substance produced by other microorganisms, e.g. penicillin; a natural plant product, e.g. quinine; or a synthetic chemical, e.g. sulphonamides.

Antifolate: A class of antibiotic, including trimethoprim, that acts against bacteria and protozoa by inhibiting dihydrofolate reductase in the folate pathway.

Arthropod: An invertebrate with jointed legs. Many insects bite humans and transmit viruses, bacteria, protozoan and metazoan pathogens to humans, e.g. *Aedes aegypti* transmits yellow fever.

Attenuation: The process of inducing a lower level of virulence in a microorganism so that it may used to induce immunity without causing disease.

Bacteriocin: A protein antibiotic substance made by bacteria to inhibit other organisms sharing the same ecological niche.

Bacteriophage: A virus that infects a bacterium. It may have a stable relationship or may cause lysis of the organism. Phages may give bacteria virulence determinants, e.g. diphtheria toxin. The pattern of phage lysis can be used to type organisms of epidemiological importance, e.g. methicillin-resistant *Staphylococcus aureus*.

Binomial: The conventional way of writing the biological name of an organism. By convention the genus and species name are written in italics, e.g. *Staphylococcus aureus*. If the genus name only is used the abbreviation sp. or spp. (single species and multispecies genera, respectively) is used. The abbreviation is written in roman characters. Common names are written in roman characters, e.g. streptococci.

Biotyping: Differentiation of organisms of the same species into clinically useful groups by measurement of physical characteristics, e.g. sugar fermentation. Biotyping is used to follow the spread of infection or to define pathogenic strains.

Capnophile: An organism that requires carbon dioxide for growth, e.g. some biotypes of *Brucella abortus*.

Capsid: Protein subunit component of virus providing the structure.

Capsule: Loose layer of material surrounding a bacterium or fungus. Usually polysaccharide but occasionally protein or lipid. Usually acts to inhibit phagocytosis by polymorphs or macrophages.

Cephalosporins: A class of antibiotics closely related to penicillins naturally more stable to degradative enzymes known as β-lactamases.

Cestode: Flatworm or platyhelminth. Includes pathogens such as tapeworm and hydatid disease.

Commensal: A microorganism that is found in the human host but does not cause disease or rarely causes disease.

Computerized tomography (CT): Computerized X-ray method to give a whole body image.

Congenital infection: An infection transmitted from mother to child *in utero*.

Conjugation: The process whereby plasmids are transmitted from one bacterium to another.

Definitive host: The definitive host is the phase of the life-cycle where the organism undergoes its sexual stage (e.g. adult stage of schistosomes).

Dermatophyte: Filamentous fungus causing infections of the skin.

Disinfection: The process of reducing pathogenic organisms to safe levels. A disinfectant is a chemical substance that is used to achieve this.

Enzyme immunoassay (EIA): A method for detecting the presence of antibodies or antigens which can be used to diagnose infectious disease or detect the concentrations of antibiotics.

Exanthema: An infection of childhood characterized by a rash, e.g. measles or chickenpox.

Family: A grouping of genera of microorganisms, e.g. the Enterobacteriaceae which includes *Escherichia coli*, *Proteus* spp., *Klebsiella* spp. and others. Family names are usually written in roman type.

Feral: Animals which are wild are described as feral.

Fimbria: Filamentous protein organelle that allows bacteria to adhere to host cell surfaces. A pathogenicity determinant for many microorganisms, e.g. *Neisseria gonorrhoeae*.

Flagellum: Hair-like organ of motility for bacteria and protozoa.

Fomite: An inanimate object that may harbour microorganisms allowing their transmission, e.g. *Pseudomonas aeruginosa* in ventilator humidifier.

Gangrene: A rapidly progressive anaerobic infection with necrosis spreading in the tissues, often caused by *Clostridium perfringens* or closely related organisms.

Genome: The genetic material of an organism.

Genus: A group of closely related organisms, e.g. *Streptococcus*. The genus is subdivided into species, e.g. *Streptococcus pyogenes* and *Streptococcus pneumoniae*.

Glycopeptides: A class of antibiotic, which includes vancomycin and teicoplanin, that acts against Gram-positive bacteria by interfering with cell wall synthesis.

Helminth: From the Greek 'helmins' meaning worm. Includes nematodes, cestodes and flukes.

Hypha: The finger-like branching structure that some fungal species form while growing, e.g. *Aspergillus fumigatus*, *Trichophyton mentagrophytes*.

Infection: Invasion of the host that causes clinical disease.

Intermediate host: The stage in the life-cycle where the pathogen undergoes asexual reproduction (trophozoites of malaria).

Involucrum: New bone found as part of the process of osteomyelitis.

Isolation: The process of separating infected from uninfected patients by providing specific facilities and practice protocols to prevent transmission of organisms in the hospital environment or community. A term also used to describe the process whereby bacteria, fungi or viruses are cultivated in artificial conditions in the laboratory.

β-Lactam: A name used to describe antibiotics active against bacteria through the β-lactam ring, e.g. penicillins and cephalosporins.

β-Lactamase: An enzyme produced by bacteria that breaks down the β-lactam ring inactivating it and rendering the organism resistant.

Lipopolysaccharide: A complex molecule consisting of polysaccharide and lipid found in Gram-negative organisms. It protects the organism from the action of complement and stimulates host responses in macrophages causing the release of cytokines, such as tumour necrosis factor.

Macrolides: A group of antibiotics, including erythromycin, that inhibit bacterial protein synthesis.

Magnetic resonance imaging (MRI): Used in whole body imaging to detect the presence of occult abscess.

Media: Solid or liquid materials used to grow microorganisms. Enrichment media enhance the growth of all organisms and selective media inhibit some organisms but allow the growth of others, e.g. MacConkey agar used for isolation of enteric organisms.

Metazoa: Multicellular organisms, some of which are human pathogens.

Microaerophilic: An organism that grows best in an atmosphere with decreased oxygen concentration.

Minimum bactericidal concentration (MBC): An objective measure of the ability of an antibiotic to kill microorganisms. It is defined as the lowest concentration of antibiotic that produces a 99.9% kill.

Minimum inhibitory concentration (MIC): This is an objective measure of the capacity of an antibiotic to inhibit the growth of a microorganism. It is defined as the lowest concentration of an antibiotic that completely inhibits growth.

Nematode: Roundworm, some of which are human pathogens, e.g. hookworms.

Nucleoside analogues: A class of antiviral agents that inhibit viruses by being incorporated into viral nucleic acid, inhibiting replication.

Nucleoside inhibitors: A class of nucleoside antiviral agents that inhibit HIV by interfering with reverse transcriptase.

Pathogenicity: The characteristic of an organism that allows it to cause human disease.

Penicillin: An antibiotic that inhibits bacteria by interfering with cross-linking in the bacterial peptidoglycan. There are many different penicillins with different pharmacokinetic properties and spectra of activity, e.g. benzylpenicillin, piperacillin. The active component is the β-lactam ring, and this name is also used to describe the class of antibiotics.

Peptidoglycan: A polysaccharide molecule that is the main structural component of bacteria.

Pili: A synonym for fimbriae.

Plasmid: A small portion of DNA found in some bacteria inherited separately from the chromosome. It may carry virulence determinants, antibiotic resistance and other genes. Plasmids permit antibiotic resistance to be rapidly transmitted between organisms.

Polymerase chain reaction (PCR): This is a method for amplifying DNA that can be used to detect the presence of pathogens in clinical specimens.

Prion: A prion is a protein that appears to act as an infectious agent. Prions are responsible for the chronic human spongiform encephalopathies including variant Creutzfeld–Jakob encephalopathy.

Prokaryote: A single-celled organism with a single chromosome made up of double-stranded DNA.

Prophylaxis: An action taken to prevent disease, e.g. vaccination or administration of antibiotics.

Protease inhibitors: A class of antiviral compounds that act by inhibiting virally encoded proteases.

Protozoa: Single-celled eukaryotic organisms, some of which are human pathogens or commensals.

Quinolones: A group of antibiotics that inhibit bacterial growth by interfering with DNA gyrase, the enzyme responsible for supercoiling bacterial DNA.

Radioimmunoassay (RIA): A serological technique that uses radioactivity to detect antibody–antigen binding.

Resistance: Property of an organism which implies that normal doses of an antibiotic will be unsuccessful. Intermediate resistance means that increased doses of an antibiotic may be successful. An equivalent term is intermediate susceptibility.

Retrovirus: A negative strand RNA virus possessing reverse transcriptase. This transcribes viral RNA into cDNA that is incorporated into the host DNA and from which further copies of the virus are made. HIV is the most important example.

RNA viruses: RNA viruses have RNA as the genetic material. It can be positive sense, i.e. can act as messenger RNA; negative sense which means it must be transcribed before acting as messenger RNA; or retrovirus which requires the action of reverse transcriptase to translate the RNA into DNA.

Saprophyte: An organism living on decaying material. Such organisms are rarely human pathogens.

Sensitivity: Alternative term to susceptibility. The term can be applied to diagnostic tests where it represents the ability to detect all of the individuals with a target disease, i.e. the true positives.

Sequestrum: Old bone found in the centre of osteomyelitis.

Southern blotting: A process of hybridization between molecules of homologous DNA. It is used in microbiology to identify the presence of identical sequences and can be used for specific diagnosis in specimens, in taxonomic investigations and in epidemiological typing.

Specificity: The ability of a diagnostic test to identify individuals who are not suffering from a target disease, i.e. the true negatives.

Spore: Specialized organelle for survival of bacteria in the environment.

Sterilization: The process of rendering an object free of living organisms.

Sulphonamides: A class of antibiotic that is active against bacteria and protozoa by inhibiting the dihydropterate synthase enzyme in the folate pathway.

Susceptibility: Property of an organism which implies that normal doses of an antibiotic will be effective. Intermediate susceptible organisms should respond to an increased dose of antibiotic. Sensitivity is an alternative term.

Susceptibility discs: Small paper discs impregnated with antibiotics used for determination of antibiotic susceptibility.

Sylvatic reservoir: A sylvatic reservoir refers to the circulation of an infectious disease in a forest environment that can spread to humans, e.g. yellow fever or Ebola virus infection.

Symmetry: Description of arrangement of virus structure, e.g. icosahedral or helical.

Teichoic acid: A structural polysaccharide component of Gram-positive bacteria.

Tetracyclines: A class of antibiotics that inhibit bacteria by interfering with protein synthesis.

Therapeutic index: The ratio of the dosage necessary for effective use and the dosage at which adverse events occur.

Toxin: A substance which interferes with biological function usually produced by a microorganism. Endotoxin is a lipopolysaccharide forming part of the cell wall of Gram-negative organisms. Exotoxins are bacterial protein antigens, e.g. cholera toxin of *Vibrio cholerae* which causes secretory diarrhoea.

Toxoid: An altered form of a toxin that has no toxic action but can induce immunity when used in a vaccine.

Transformation: A process whereby bacteria take up naked DNA and incorporate it into their genome.

Transposon: A moveable genetic element that is able to encode its own transmission. It may carry antibiotic resistance genes.

Trematode: Flukes. These organisms can be located in the blood vessels, e.g. schistosomes; the lung, paragonimiasis; or the liver, fascioliasis.

Trophozoite: The growing stage of some protozoan pathogens, such as *Plasmodium falciparum*.

Typing: A process of dividing microorganisms into groups so that the spread of individual organisms can be followed in the hospital environment or the community.

Vaccination: The process of administering a substance to induce immunity to an infectious agent.

Vector: A living vehicle of transmission of infectious disease. The *Anopheles* mosquito is the vector of malaria.

Virulence: Characteristic of a microorganism that allows it to cause severe disease, e.g. streptococcal erythrogenic toxin.

Western blotting: A method to detect the presence of specific antibodies that is sometimes used to diagnose infectious diseases, such as HIV.

Zoonosis: An animal disease that can be transmitted to humans, e.g. brucellosis.

Antibiotic table

Antibiotic	Route	Related antibiotics	Mechanism of action	Spectrum of activity
Benzylpenicillin (β-lactam)	P		Inhibits peptidoglycan cross-linking	Streptococci and oral anaerobes, *Neisseria*
Phenoxymethylpenicillin (β-lactam)	Oral		Inhibits peptidoglycan cross-linking	Streptococci and oral anaerobes, *Neisseria*
Ampicillin (β-lactam)	O/P	Pivampicillin, amoxicillin	Inhibits peptidoglycan cross-linking	Streptococci and some Enterobacteriaceae
Flucloxacillin (β-lactam)	O/P	Methicillin	Inhibits peptidoglycan cross-linking	*S. aureus* and streptococci
Ticarcillin (β-lactam)	P	Carbenicillin	Inhibits peptidoglycan cross-linking	Enterobacteriaceae and *Pseudomonas*
Azlocillin (β-lactam)	P	Piperacillin	Inhibits peptidoglycan cross-linking	Enterobacteriaceae and *Pseudomonas*
Clavulanic acid (β-lactam)	P/O	Tazobactam	β-lactamase inhibitor	No intrinsic activity
Aztreonam (monobactam)	P		Inhibits peptidoglycan cross-linking	Gram-negative bacteria only
Meropenem (carbapenem)	P	Imipenem	Inhibits peptidoglycan cross-linking	Gram-positive, Gram-negative and obligate anaerobes
Cefradine (cephalosporin)	O/P	Cefaclor, cefalexin	Inhibits peptidoglycan cross-linking	Gram-positive cocci and some Enterobacteriaceae
Cefixime (cephalosporin)	O	Cefpodoxime	Inhibits peptidoglycan cross-linking	Gram-positive cocci, not staphylococci and Enterobacteriaceae
Cefotaxime (cephalosporin)	P	Ceftriaxone, cefuroxime	Inhibits peptidoglycan cross-linking	Gram-positive cocci and Enterobacteriaceae
Ceftazidime (cephalosporin)	P		Inhibits peptidoglycan cross-linking	Enterobacteriaceae and *Pseudomonas*
Tetracycline	O/P	Oxytetracycyline, doxycycline, minocycline	Inhibits protein synthesis	Gram-positive and -negative bacteria, *Chlamydia*, *Mycoplasma*, *Riskettsia*, some protozoa
Gentamicin (aminoglycoside)	P	Netilmicin, amikacin, streptomycin, tobramycin	Inhibits protein synthesis	Staphylococci, Enterobacteriaceae, mycobacteria, *Pseudomonas*
Erythromycin (macrolide)	O/P	Azithromycin, clarithromycin	Inhibits protein synthesis	Gram-positive cocci, *Haemophilus*, some anaerobes, *Mycoplasma*, *Chlamydia*, *Rickettsia*
Clindamycin (licosamide)	O/P		Inhibits protein synthesis	Gram-positive cocci, obligate anaerobes, *Toxoplasma*, some *Plasmodium*
Chloramphenicol	O/P		Inhibits protein synthesis	Gram-positive cocci, obligate anaerobes, *Haemophilus*, some Enterobacteriaceae, *Chlamydia*, *Mycoplasma*, *Rickettsia*
Fusidic acid	O/P		Inhibits protein synthesis	Gram-positive cocci but not streptococci, *Neisseria* and *Haemophilus*

Continued p. 114

Antibiotic	Route	Related antibiotics	Mechanism of action	Spectrum of activity
Vancomycin (glycopeptide)	P	Teicoplanin	Inhibits peptidoglycan synthesis	Gram-positive bacteria only
Co-trimoxazole	O/P		Antifolate	*Pneumocystis*
Trimethoprim	O/P		Antifolate	*Haemophilus*, some Gram-negative bacteria and Gram-positive cocci
Rifampicin (rifamycin)	O	Rifabutin	Inhibits RNA polymerase	Mycobacteria, *Chlamydia*, *Mycoplasma*, *Riskettsia*, Gram-positive cocci and Gram-negative cocci
Isoniazid	O		Inhibits cell wall synthesis	Mycobacteria only
Ethambutol	O		Uncertain	Mycobacteria only
Pyrazinamide	O		Uncertain	Mycobacteria only
Metronidazole (nitroimidazole)	O/P	Tinidazole	Damages microbial DNA	Obligate anaerobes, *Trichomonas*, *Giardia* and amoebae
Ciprofloxacin (fluoroquinolone)	O/P	Ofloxacin, levofloxacin	Inhibits DNA supercoiling	Enterobacteriaceae, *Pseudomonas*, some Gram-positives
Nitrofurantoin (nitrofuran)	O		Damages DNA	Urinary Gram-negative bacteria
Fluconazole (imidazole)	O/P	Itraconazole	Inhibition of ergosterol metabolism	Yeasts
5-Flucytosine	O/P		Nucleoside inhibitor	*Candida*
Amphotericin (polyene)	P	Nystatin	Pores in plasma membrane	Yeasts and filamentous fungi
Aciclovir (guanosine analogue)	O/P	Valaciclovir, famciclovir	Nucleoside inhibitor	Herpes simplex and VZV
Amantadine (tricyclic amine)	O		Prevents viral uptake	Influenza A
Zidovudine (nucleoside reverse transcriptase inhibitor)	O	Didanosine, lamivudine, stavudine, zalcitabine	Nucleoside reverse transcriptase inhibitor	HIV-1
Indinavir (protease inhibitor)	O	Ritonavir, saquinavir	Protease inhibitor	HIV-1
Nevirapine	O	Delavirdine	Non-nucleoside reverse transcriptase inhibitor	HIV-1
Ganciclovir (guanosine analogue)	O/P		Nucleoside inhibitor	CMV
Foscarnet (pyrophosphate analogue)	P		Nucleoside inhibitor	CMV
Ribavirin (synthetic nucleoside)	Aerosol		Nucleoside inhibitor	RNA viruses
Chloroquine (4-amino-quinolone)	O		Uncertain	Malaria, amoebae
Mefloquine (quinoline-methanol)	O		Uncertain	Malaria

Continued p. 115

Antibiotic	Route	Related antibiotics	Mechanism of action	Spectrum of activity
Quinine (cinchona alkaloid)	O/P		Uncertain	Malaria
Pentamidine (diamidine)	P		Inhibits polyamine metabolism	*Pneumocystis, Leishmania*
Mebendazole (benzimidazole)	O		Inhibition of β-tubulin	Intestinal nematodes
Praziquantel (pyrazino-quinolone)	O		Interferes with calcium homeostasis	*Schistosoma, Taenia* and other cestodes
Ivermectin (avermectin)	O		GABA antagonist	Filaria including *Onchocerca*
Piperazine (heterocyclic organic base)	O		Worm paralysis	Intestinal nematodes, including threadworm

Abbreviations: CMV, cytomegalovirus; GABA, γ-aminobutyric acid; O, oral; P, parenteral; VZV, varicella-zoster virus.

Index

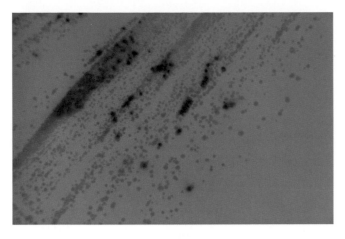

1 The oxidase test. The plate has been flooded with the oxidase reagent and colonies have rapidly turned purple. This test is used in the laboratory to help identify genera such as *Neisseria* and *Pseudomonas*.

2 Coagulase. Demonstration of coagulase enzyme is the principal way of differentiating *Staphylococcus aureus* from other staphylococci. In this picture one of the tubes has coagulated and so the meniscus does not move when the tube is tipped compared with the other tube.

3 The catalase test is a simple method to demonstrate the presence of catalase. Colonies are picked up by a wooden stick and placed in some hydrogen peroxide. If the colony possesses catalase bubbles of oxygen will be given off rapidly. Catalase is used to distinguish between organisms, for example staphylococci are catalase positive but streptococci are catalase negative.

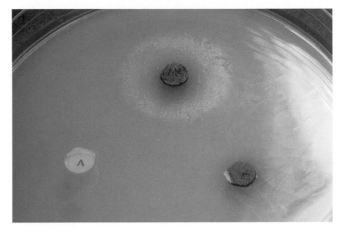

4 XV discs. *Haemophilus influenzae* is dependent on two factors: haemin and NAD (also known as X and V). When cultivated on nutrient agar the organism is only able to grow if both factors are present. This is illustrated in this figure because growth only occurs around the XV disc.

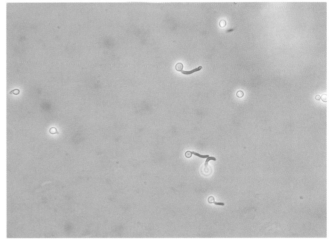

5 Germ tube. *Candida albicans* can be differentiated from other species that are usually less invasive using the germ tube test. The organism is incubated in serum for a few hours and if it is *Candida albicans*, the small projection illustrated here, will develop.

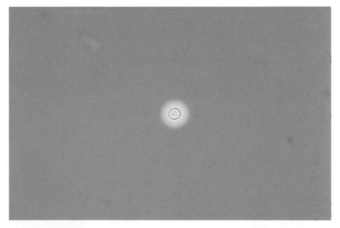

6 *Cryptococcus.* *Cryptococcus neoformans* is surrounded by a capsule. It is visualized in an indian ink preparation by the halo shown here.

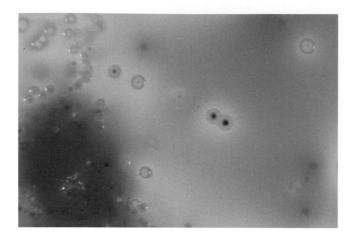

7 Xylose lysine desoxycholate medium is commonly used to isolate enteric pathogens. Coliform organisms appear as yellow colonies and potential pathogens, e.g. *Salmonella* and *Shigella* appear as pale colonies. Those organisms that produce H₂S (as some *Salmonella* do) may have a black centre, as seen here.

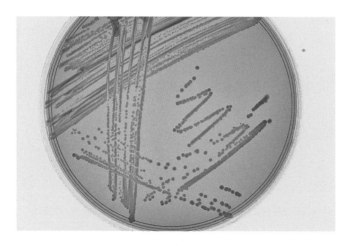

8 MacConkey medium. MacConkey agar contains bile salts, lactose and an indicator. It is selective for enteric organisms because only organisms that tolerate bile salts will grow. Organisms that ferment lactose produce acid that turns the indicator (and the colonies) red. In this figure *E. coli* is illustrated.

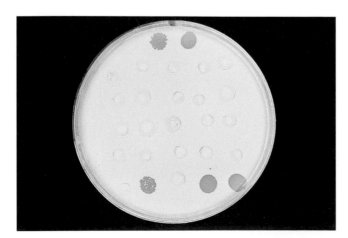

9 Phage typing. *Staphylococcus aureus* are traditionally typed by an internationally agreed set of phages. Each strain differs in the phages that will lyse the organisms leaving a plaque as seen here. Different organisms will have different patterns of lysis. Molecular techniques such as pulse field gel electrophoresis are now also used to differentiate strains of *Staphylococcus aureus*.

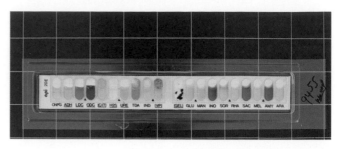

10 Biochemical testing of Enterobacteriaceae. Many organisms require a series of biochemical tests for identification. This strip method is one of many that is used to speed this process in the laboratory. This example is used to identify the Enterobacteriaceae. The results of the tests are given a code and the code is compared against tables produced by the manufacturers. Automated-computerized versions of this approach are increasingly being used in laboratories.

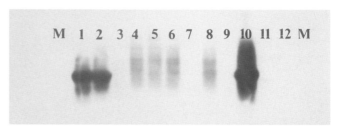

11 Southern blotting. An important method of ensuring that the PCR product obtained in a reaction is the correct one is to used labelled DNA in a Southern blot. Labelled DNA is hybridized to PCR products; if there is homology between the probe and the PCR product hybridization will take place. The figure illustrates a TB PCR.

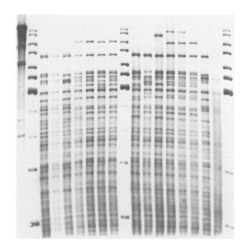

12 Molecular typing. Many bacterial pathogens are typed using restriction fragment linked polymorphism methods. DNA is taken from the organism and digested by an endonuclease. The products are run on a gel and their position demonstrated by a hybridization step. The differences in the patterns can be used to distinguish different strains or suggest that some strains are identical. This is of use in following outbreaks of organisms in hospital or the community.

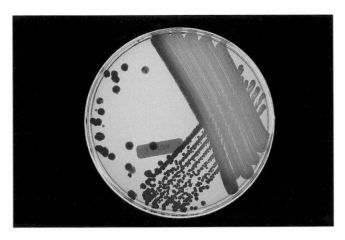

13 *Serratia marcescens*. This organism is one of the Enterobacteriaceae. They normally produce colourless colonies. *Serratia marcescens* is an opportunistic pathogen that sometimes produces bright red pigment.

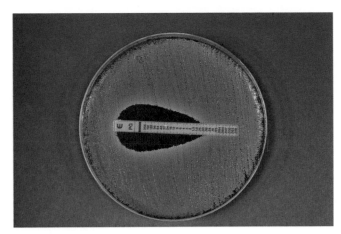

16 Susceptibility testing by E-test. The E-test consists of a strip impregnated with differing concentrations of antibiotic throughout its length. The point at which bacterial growth reaches the strip is related to the MIC of the organism which can be read from the printed scale.

14 LJ slopes. Mycobacteria are slow growing so they are incubated on medium that inhibits the growth of non-mycobacterial organisms, in this case by incorporating malachite green. The specimen is inoculated onto the slope of a tube with a tight fitting lid to prevent it drying out over the long incubation periods required for isolation.

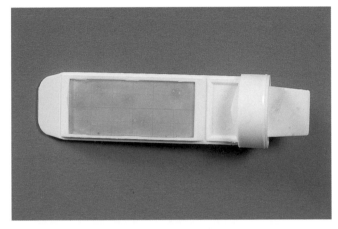

17 Dipslide. The dipslide is a way of collecting and transporting urine specimens to the laboratory. The dipslide is dipped into the urine and the excess allowed to drain. The dipslide is then placed in its plastic container and transported to the laboratory for incubation. The number of colonies can be used to determine whether the organisms are present in significant numbers.

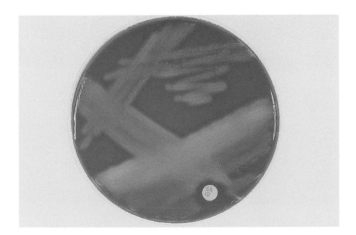

15 β-Haemolytic streptococci. Streptococci are separated by their ability to haemolyse blood. Organisms that cause complete haemolysis such as *Streptococcus pyogenes*, as in this figure, are called β-haemolytic. Where partial haemolysis occurs this is called α-haemolysis.

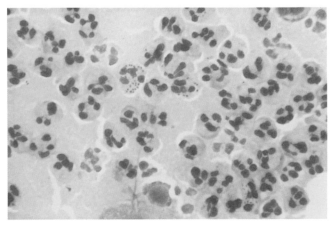

18 CSF Gram stain. A large number of polymorphonuclear leucocytes are seen along with small Gram-negative diplococci. This patient is suffering from meningitis caused by *Neisseria meningitidis*.

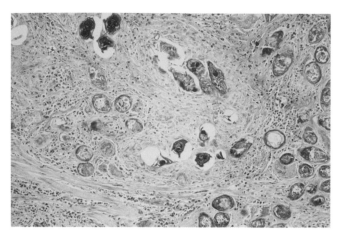

19 *Schistosoma* in tissue. The eggs of *Schistosoma* are expelled into the urine or faeces. Eggs that remain in the tissues cause intense inflammation and subsequently fibrosis. It is this that is responsible for the later complications of schistosomiasis. In this illustration many eggs can be seen in a cervical biopsy.

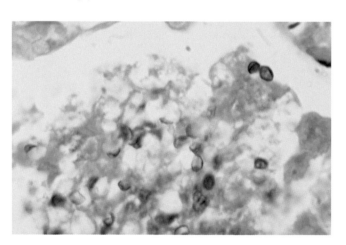

20 *Pneumocystis*. *Pneumocystis carinii* infection occurs among patients with a severe deficit in T cell function. It became a common condition among HIV-infected individuals in whom the CD4 count had fallen below 0.2×10^9/litre.

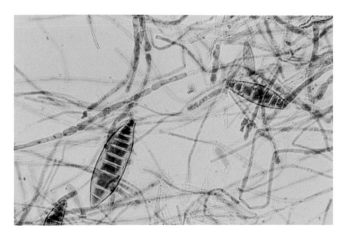

21 *Microsporum*. Dermatophytes are identified by the characteristic morphology of their macroconidia. This organism is typical of *Microsporum canis*.

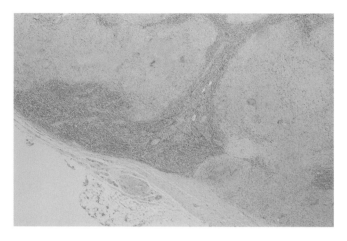

22 The tuberculosis granuloma is characterized by the presence of many round cells, caseous necrosis and Langhans giant cells. A few mycobacteria can sometimes be visualized by Ziehl–Nielsen staining.

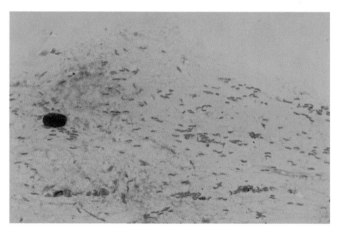

23 Gram stain from gastric mucosal biopsy of a patient with a past history of duodenal ulcer. Gram-negative curved and spiral bacilli typical of *Helicobacter pylori* are seen. The histology showed chronic active gastritis.

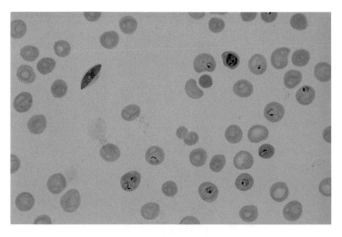

24 Blood film with *Plasmodium falciparum*. This blood film is taken from a patient with severe *P. falciparum* malaria. There are very many ring-shaped merozoites present, and a typical gametocyte can also be seen. There is a single early schizont. This is not normally seen in patients with *P. falciparum* malaria as they are usually adherent to capillary endothelium.

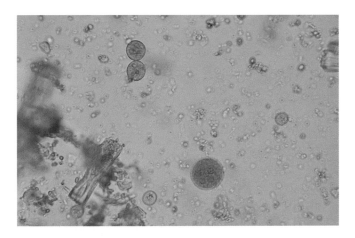

25 Amoebic cysts in stool. This is a sample of stool and two different species of *Entamoeba* are shown. The larger is *Entamoeba coli* and the smaller cysts are *E. histolytica* or *E. dispar*. Differentiation of these requires specialized tests.

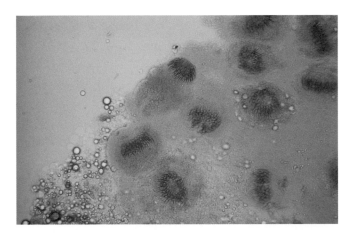

26 *Echinococcus* protoscolices. In hydatid disease there is a cyst that is usually found in the liver. Inside the cyst are protoscolices and these are illustrated here.

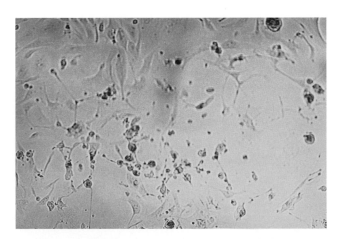

27 Cytopathic effect. The growth of viruses inside cell lines can be recognized by the cytopathic effect. In this figure enterovirus are shown and they cause rounding, shrinking and loss of contiguousness in the cell monolayer.

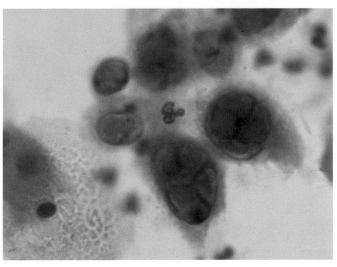

28 Herpetic vesicle. Microscopic appearance of cells from a herpetic vesicle. The cells typically have intranuclear inclusion bodies.

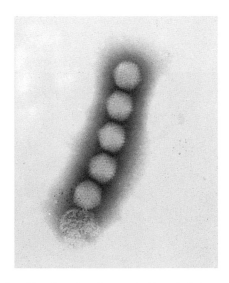

29 Adenovirus. The adenovirus shows typical icosahedral symmetry.

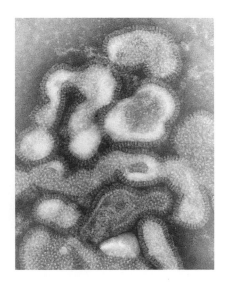

30 Influenza virus. The influenza virus is an enveloped RNA virus. The haemagglutinin and neuraminidase proteins expressed as 10 nm spikes on the envelope can be seen here.

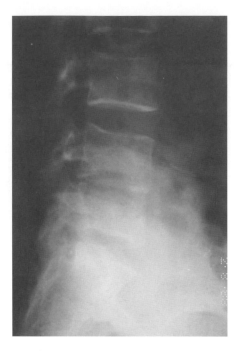

31 Tuberculosis of the spine. In this X-ray of the lumbar spine there is irregularity and loss of disc space and there is radiolucency in two of the vertebrae. This film was from a 23-year-old African student who presented with chronic back pain. He had a strong positive Mantoux reaction and the diagnosis was tuberculosis of the spine.

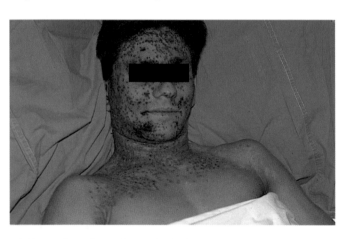

32 Eczema herpeticum. Herpes simplex is an important cause of acute skin ulceration. It may relapse later in life to cause cold sores. Patients with eczema are especially susceptible to herpes infection which can spread more widely, as in this case.

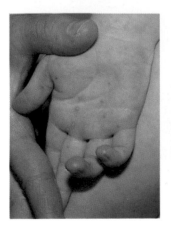

33 Hand, foot and mouth. The human parvovirus B19 causes hand, foot and mouth disease and the typical lesions are illustrated here.

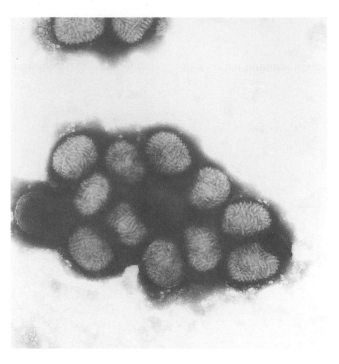

34 Poxvirus. Under EM the poxviruses resemble balls of tightly wound wool. This virus is taken from a patient with molluscum contagiosum.

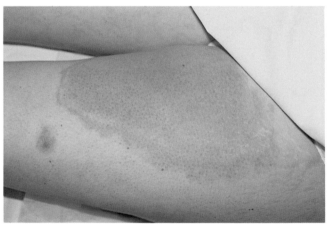

35 Erysipelas. Erysipelas usually affects the skin of the face but can affect other parts of the body as in this case. It is an intradermal infection usually with *S. pyogenes*. The edges of the lesions are well demarcated and the affected skin feels hot to the touch.

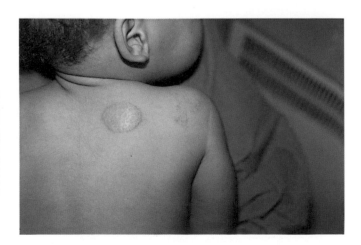

36 Tinea corporis. Infection of the skin with dermatophyte fungi is known as ringworm. The lesions typically have a red edge with a central area of more normal skin as in this child.

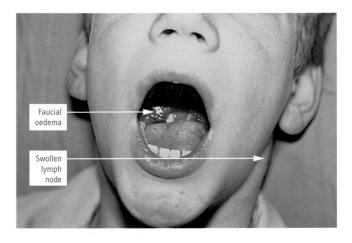

37 Tonsillitis. *Streptococcus pyogenes* is the most common cause of bacterial pharyngitis. It may be complicated by peritonsillar abscess, or later by glomerulonephritis or rheumatic fever.

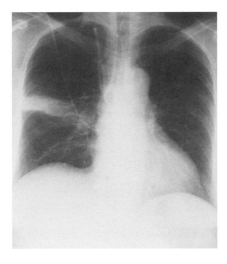

38 Lobar pneumonia. This patient with non-Hodgkin's lymphoma presented with fever, abdominal discomfort and confusion. Three days later respiratory signs and new changes developed on chest X-ray. A urine antigen test for *Legionella* was positive. The patient had recently travelled to Spain.

39 Pseudomembranous colitis. Pseudomembranous colitis is characterized by intense inflammation erupting from below the muscularis mucosae producing an accumulation of necrotic debris.

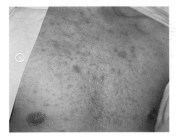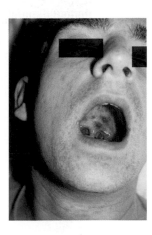

40 HIV seroconversion illness. The initial stage of HIV is often characterized by a seroconversion illness. This is a febrile illness occurring 5–8 weeks after infection persisting for 2–3 weeks. It is associated with lymphadenopathy and sometimes a rash as in the case illustrated here.

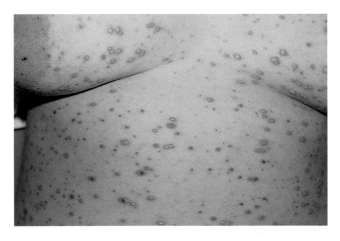

41 Chickenpox. The rash of chickenpox has lesions at different stages of development known as cropping. Lesions occur most densely on the trunk.

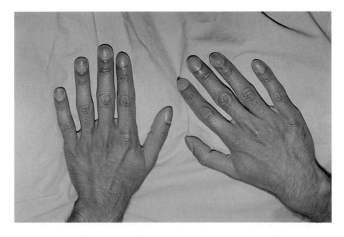

42 Finger clubbing in endocarditis. This patient had a 3-month history of general malaise and a low grade fever. On examination of his heart a murmur could be heard and his fingers show evidence of finger clubbing suggesting the diagnosis of endocarditis.

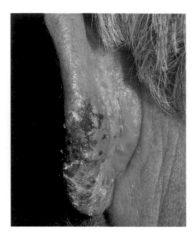

43 *Leishmania.* This man presented with a chronic infection of his pinna. He had been working in the desert on an oil pipeline with a crew of other expatriates. Many of them had a similar lesion. This is caused by *Leishmania* and the chronic granulomatous reaction to it in the skin.

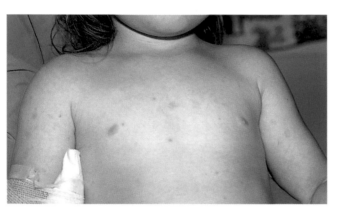

44 Acute meningococcal rash. This child has only been unwell for a few hours. The typical petechiae of meningococcal septicaemia can be seen and will progress rapidly unless benzylpenicillin is given urgently.

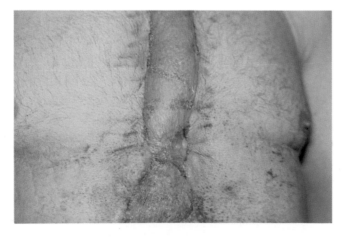

45 Methicillin-resistant *Staphylococcus aureus* (MRSA). MRSA are transmitted in the hospital environment. They may infect patients who are already compromised by their disease. In this case the patient's sternotomy wound has become infected with MRSA.

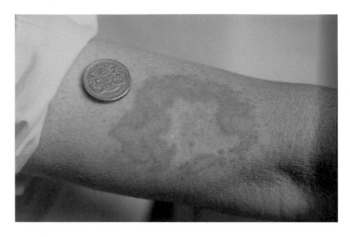

46 Erythema chronicum migrans. Up to 80% of patients develop a rash around the area of the tick bite. It takes the form of an expanding ring of erythema with central clearing. Multiple lesions are occasionally seen.

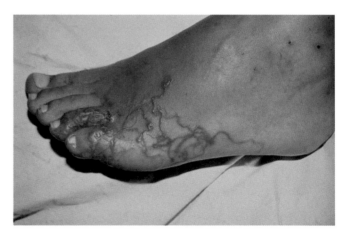

47 Cutaneous larva migrans. This patient presented with the rash shown here that is intensely itchy. The diagnosis is cutaneous larva migrans caused by a dog hookworm. It is unable to complete its life-cycle in the human host but remains in the skin to produce this rash.

The authors gratefully acknowledge the assistance of Dr Julie Crow, Mr Devendra Kothari, Ms Linda Tilling, Dr Rob Gargan and Mrs Ann Smith for their help in originating these illustrations. The authors are especially grateful to Dr Bannister for permission to use figures from *Infectious Diseases*, Bannister, Begg & Gillespie, Blackwell Science, 1997.